What's That You Have In Your Hands? ~ 1

What's That You Have In Your Hands?

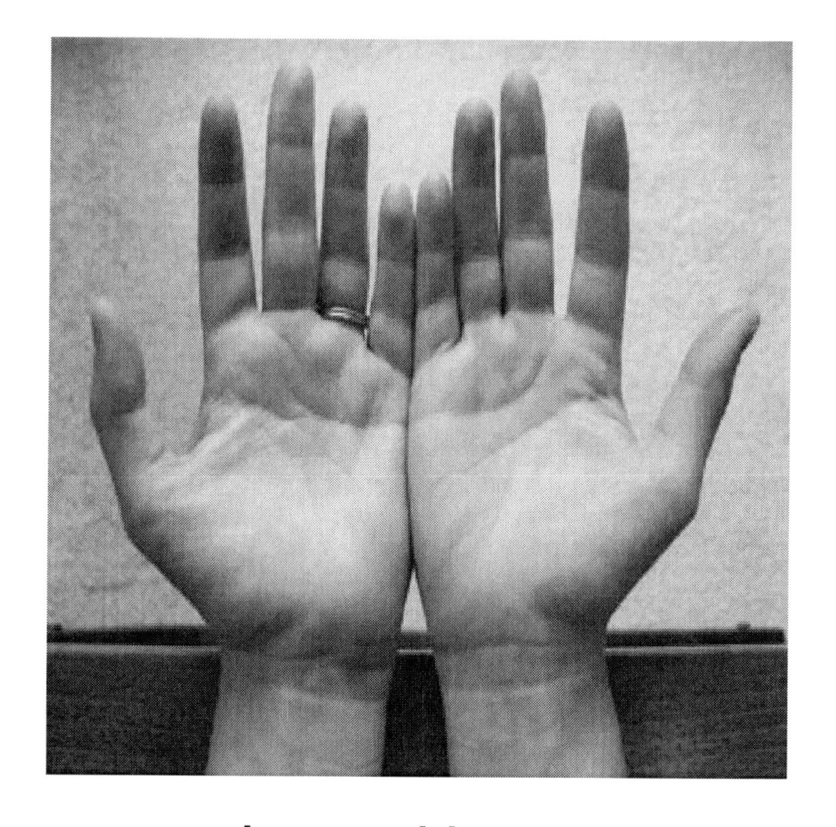

Jeanne Halsey

What's That You Have In Your Hands?

By **Jeanne Halsey**

ISBN 978-0-557-42704-8

Unless otherwise indicated, all Scriptures are quoted from *The New American Standard Bible,* Open Bible Expanded Edition.

For more information, contact:

ReJoyce Books
4424 Castlerock Drive
Blaine, Washington 98230
United States of America

www.halseywrite.com

Table of Contents

Acknowledgements

Writing this book was a delightful experience, definitely a strong moment in my writing career. It mostly occurred during the month of May 1999, and almost entirely during a two-week period when my husband **Kenneth Halsey** and I traveled to Barbados, West Indies, to celebrate our 25th wedding anniversary.

Kenneth was an avid runner, and being on a "honeymoon vacation" did not deter him from continuing his training to possibly run in another Marathon race that year, so while he was up early every morning sweating out his multi-miles in the balmy Bajan climate ... I was scribbling in a little spiral-bound notebook, while sitting on the balcony of our hotel room. (Although I eventually varied my locations, including some on the pure white "sugar" sand at famed Accra Beach, or on a lounger next to the hotel's fresh-water pool.)

I did not use my typical computerized method of writing, which enables me to write and re-write with ease – instead I literally ran the ink out of a new black ballpoint pen while "watching" Holy Spirit put words and phrases and entirely new thoughts onto the paper. I will always treasure that little notebook, with its ocean-water, pool-water, sunscreen, pineapple juice, and sweat stains that made the pages both limp and crisp; with the part where I had to switch to a blue pen; and my "shorthand" and Scripture references. A lot of creativity went into that little book.

There are special friends of whom I want to make a special note; they helped provide the restful, creative locations wherein I wrote this book. Two brothers, **Simon and Chris Walker,** who, along with their mother **Ann**

Walker, are a family that owns and operates the Oasis Hotel in Rockley, Christ Church, Barbados. They are fine Spirit-filled Christians and generous-hearted people ... and they run a wonderful little hotel that's just a bit off the main thoroughfare, with just enough remoteness to allow birds' singing to overcome island-traffic sounds. They have a beautiful garden-like setting with rioting flowers and stately tropical trees, and an atmosphere of supreme relaxation that refreshes the body and renews the soul. If you visit Barbados, I recommend their fellowship – and their hotel – to you.

There is another person whom I have never met but to whom I owe a debt of inspiration. **Helena Barrington** wrote the song, *"That's When,"* which I was introduced to on **Alvin Slaughter's** CD, *"God Can!"* This song reverberated in my spirit for several years until finally this book emerged, with its flavor added. *"Thank you, Helena, for putting into music those thoughts that God had long been putting into my own heart."*

There was also the Adult Sunday School Class at a little church in Boyle, Alberta, thousands of miles north of balmy Barbados. Early in 1999, I had attended a three-day Women's Conference where my sisters (the late) **Judy Gossett** and **Reba Rambo-McGuire** were the main speakers. I was asked by the pastor of the host-church to bring a brief lesson for the adults during the Sunday School hour. On the Saturday night, I sat in my hotel room and ran through my electronic concordance, and eventually came up with the basic outline on the subject of "hands," which I brought to the class the next morning. I don't know if they got very much out of my rough teaching, but what God was saying certainly struck my heart and continued to speak to me, until I finally decided to breathe more life into thus subject, making it into a full-length book.

There have also were some rather unusual things about this book: several times, I cited whole songs; I have quoted other authors extensively; I included drama (or comedy) scripts; to illustrate certain points, I even borrowed from other books I have previously written. I gave myself permission to try all these options ... and it certainly made for your "not run-of-the-mill manuscript." I hope the Reader doesn't get lost in the varying transitions between writing styles.

Dedication

I have a lifetime friend who is a living example of allowing God to take what she has in her hands and help her make it grow into an answer to her dilemma. **Grace Elizabeth Joy Anderson Graham** is the godmother of our two children, and has walked through many of Life's challenges and triumphs with our family, remaining a good friend and a Christian sister. She is now almost literally that widow of 1 King 17: her husband (our very good friend) George Graham died at age 58 on January 4, 1999, leaving her with precious memories but very little monetarily.

Then Grace, who has always been a psalmist and a Gospel-style pianist, looked to her hands – through which the music from her heart flows – and offered her songs to God for His blessing, and as a means of provision for her life without George. I believe her faith and response to *"What's that you have in your hands?"* will see her through. *"Grace, I pray God's richest blessings on you as you continue to trust in Him, to serve Him with your talents and abilities, and to minister to the Body of Christ and the world through your music. May your talented hands play beautiful music, and may your songs bring great glory to Him."*

For several years, I walked with my younger brother **Donnie Gossett** through an unusual and difficult path. Some of our extensive conversations about Life contributed to the tone of this book. *"In this world of card-carrying Christian cynics, I still encourage you to seek – not so much the hand of God – but the face of God ... and I believe He will then say to you, 'What's that you have in your hands, Donnie?' You will find strange and marvelous ways to achieve more of the greatness which He has* *already placed in you!"*

As always, I appreciate my wonderful husband, R. Kenneth A.J. Halsey, with whom I have walked through 36 years of vibrant living. *"I like you even more than I did when I first met you ... and I love you even more, with all the strength and depth of our shared joys and sorrows. Here's to the next 36 years together!"*

Finally, this book would not be what it is without that amazing touch of my Best Friend, **Holy Spirit of Jesus Christ.** There were many times when I stopped to read back, only to be startled by what was on the pages – I genuinely could not remember writing some of those thoughts and insights! I know He had something special He wanted to say through these pages – things for specific people -- and so I was glad to "get out of His way" and let Him express the heart of the Father for His children. *"Thank You again, Heavenly Father, for giving me this gift of writing. I will continue to use my hands to bring glory to You."*

– ***Jeanne Halsey***
(revised) April 2010

Introduction

This is not a book about self-esteem, but one about how God views His children – the gifts He had placed within us that He intends to activate in our lives when we need them. I really don't like the metaphor, *"I'm at the end of my rope"* because it is too limiting, too self-important, and not enough God-acknowledging:

> *For this reason I say to you, "Do not be anxious for your life, as to what you shall eat, or what you shall drink; nor for your body, as to what you shall put on. Is not life more than food, and the body more than clothing?" Look at the birds of the air, that they do not sow, neither do they reap, nor gather into barns, and yet Your Heavenly Father feeds them. Are you not worth much more than they?*
>
> *And which of you by being anxious can add a single cubit to his life's span? And why are you anxious about clothing? Observe how the lilies of the field grow: they do not toil nor do they spin, yet I say to you that even Solomon in all his glory did not clothe himself like one of these. But if God so arrays the grass of the field, which is alive today and tomorrow is thrown into the furnace, will He not much more do so for you, O men of little faith?*
>
> *Do not be anxious then, saying, "What shall we eat?" or "What shall we drink?" or "With what shall we clothe ourselves?" For all these things the Gentiles eagerly seek; for your Heavenly Father knows that you need all these things.* ***But seek first His Kingdom and His righteousness; and ll these things shall be added to you.*** *Therefore do not be anxious for tomorrow, for*

tomorrow will care for itself. Each day has enough trouble of its own.

– Matthew 6:25-37

I firmly believe God has created us for **His** purposes – He knows our whole lives, from our beginnings to our endings, and **nothing** ever catches Him by surprise. And since He is a loving Father and has a purpose for each of our lives, why would He fail to equip us with all the tools, all the gifts, all the opportunities needed to fulfill our destinies?

More often than not, the thing that puts us in the place of feeling like we've reached the proverbial "rope's end" is usually our stubborn self-will. We make bad choices, we do stupid and selfish things, we get ourselves into trouble – and sometimes, it is just uncontrollable bad times – and **then** we cry out, *"Help me, God, I'm dying here!"*

Let us look at some of the ways God has placed our own salvation into our hands.

Chapter 1: **Moses ... In His Raw State**

Moses was raised in the lap of luxury, educated like the son of the king (for all he ever knew, he **was** the son of the king!), and had a wonderful future mapped out for him. In today's terminology, he was like the privileged son of a major tycoon, with all the resources of a wealthy corporation at his fingertips, with all the assurance of fame and fortune at his command. He was very well educated in all the known intelligence, arts and sciences, governmental skills; he was an athletic warrior, sophisticated and clever.

Then he encountered a series of "reversal of fortune" ... and ended up tending sheep for somebody else in a desolate, distant land – a nobody with nothing. He literally was not the man he thought he was, and all his plans, hopes and dreams had been destroyed forever. I imagine he felt like he was at the end of his rope ... or in this case, he had the "short end of the stick" since he was merely a lowly shepherd equipped with a shepherd's staff.

Although it is lengthy, I invite you to carefully witness this paraphrased dialogue, and see if you recognize yourself anywhere:

God Speaks to Moses

> *Now Moses was pasturing the flock of Jethro, his father-in-law, the priest of Midian; and he led the flock to the west side of the wilderness, and came to Horeb, the mountain of God. And the angel of the Lord appeared to him in a blazing fire from the midst of a bush, and he looked, and behold, the bush was burning with fire, yet the bush was not consumed. So Moses said, "I must turn aside now, and see this marvelous sight, why the*

bush is not burned up.”

When the Lord saw that he turned aside to look, God called to him from the midst of the bush, and said, “Moses, Moses!” And Moses answered, “Here I am.”

Then God said, “Do not come near here; remove your sandals from your feet, for the place on which you are standing is holy ground.” He said also, “I am the God of your father, the God of Abraham, the God of Isaac, and the God of Jacob.” Then Moses hid his face, for he was afraid to look at God.

And the Lord said, “I have surely seen the affliction of My people who are in Egypt, and have given heed to their cry because of their taskmasters, for I am aware of their sufferings. So I have come down to deliver them from the power of the Egyptians, and to bring them up from that land to a good and spacious land, to a land flowing with milk and honey. ...

“And now, behold the cry of the sons of Israel has come to Me; furthermore, I have seen the oppression with which the Egyptians are oppressing them. Therefore, come now, and I will send you to Pharaoh, so that you may bring My people, the sons of Israel, out of Egypt.”

God Answers Moses

But Moses asked God, “Who am I that I should go to Pharaoh, and that I should bring the sons of Israel out of Egypt?”

And He answered, “Certainly I will be with you, and this shall be a sign to you that it is I Who have sent you: when you have brought the people out of Egypt, you shall worship God at this mountain."

Then Moses said to God, "Behold, I am going to the sons of Israel, and I shall say to them, 'The God of your father has sent me to you.' Now they may ask me, 'What is His Name?' What shall I say to them?"

And God answered Moses, "I AM WHO I AM"; and He said, "Thus you shall say to the sons of Israel, 'I AM has sent me to you.'" ...

Moses Argues With God

Then Moses asked, "What if they will not believe me, or listen to what I say? For they may say, 'The Lord has not appeared to you.'"

***And the Lord answered him, "What is that in your hand?"** And he replied, "A staff." Then He said, "Throw it on the ground." So he threw it on the ground, and it became a serpent; and Moses fled from it.*

But the Lord said to Moses, "Stretch out your hand and grasp it by its tail" – so he stretched out his hand and caught it, and it became a staff in his hand – "that they may believe that the Lord, the God of their fathers, the God of Abraham, the God of Isaac, and the God of Jacob, has appeared to you."

And the Lord furthermore said to him, "Now put your hand into your bosom." So Moses put his hand into his bosom, and when it took it out, behold, his hand was leprous like snow. Then He said, "Put your hand into your bosom again." So he put his hand into his bosom again; and when he took it out of his bosom, behold, it was restored like the rest of his flesh.

"And it shall come about that if they will not believe you or heed the witness of the first sign, they may believe the witness of the last sign. But it shall be that if they will not believe even those two signs or heed what you say, then you shall take some water from the Nile and pour it on the dry ground; and the water which you take from the Nile will become blood on the dry ground."

Moses Still Doesn't Get It

Then Moses pleaded with the Lord, "Please, Lord, I have never been eloquent, neither recently nor in time past, nor since Thou hast spoken to Thy servant; for I am slow of speech and slow of tongue."

And the Lord said to him, "Who has made Man's mouth? Or Who makes him dumb or deaf, or seeing or blind? Is it not I, the Lord? Now then go, and I, even I, will be with your mouth, and teach you what you are to say."

But he argued, "Please, Lord, now send the message by whomever Thou wilt."

Then the anger of the Lord burned against Moses, and He said, "Is there not your brother, Aaron the Levite? I know that ***he*** *speaks fluently. And moreover, behold, he is coming out to meet you; when he sees you, he will be glad in his heart.*

"And you are to speak to him and put the words in his mouth; and I, even I, will be with your mouth and his mouth, and I will teach you what you are to do. Moreover, she shall speak for you to the people; and it shall come to pass that he shall be as a mouth for you,

and you shall be as the voice of God to him. And you shall take in your hand this staff, with which you shall perform the signs."

– Exodus 3:1-14, 4:1-17; paraphrased, emphasis added

In what did Moses believe? His inadequacies, his failures. He was more motivated by fear of Man than by fear of God. He dared to argue with God. God revealed to Moses His own **identity** (Exodus 3:6) ... His **plan** (3:8) ... His **promise** (3:12, 15). He gave Moses His **authorization** (3:14) ... and His **tools** (3:4-8). But still the argument continued, and Moses' fear/unbelief continued to be greater than his faith/belief ... until God *"burned angrily"* (4:14). Beware of the wrath of God!

There are several keys to understanding how God helps us realize our potential and activate His method of keeping us on track to fulfill His plan for our lives.

Key #1 – Look For God

Now Moses was pasturing the flock of Jethro his father-in-law, the priest of Midian, and he led the flock to the west side of the wilderness, and came to Horeb, the mountain of God. And the angel of the Lord appeared to him in a blazing fire from the midst of a bush; and he looked, and behold, the bush was burning with fire, yet the bush was not consumed. So Moses said, "I must turn aside now, and see this marvelous sight, why the bush is not burned up."

– Exodus 3:1-3

Take your focus off of yourself and your problems, and **look for the unusual,** because in the supernatural is where you will most often find God – *"turn aside and see this great sight"* (Exodus 3:3). There is value in the adage, *"Stop and* *smell the roses,"* but there is greater value in knowing that salvation is to be found in the hand of God.

Key #2 – Know Who God Is *Not*

When the Lord saw that he turned aside to look, God called to him from the midst of the bush, and said, "Moses, Moses!"

And he said, "Here I am."

Then He said, "Do not come near here; remove your sandals from your feet, for the place on which you are standing is holy ground."

– Exodus 3:4-5

Recognize that God is **not** your "peer," not just another "self-help fix" to changing your circumstances – *"for the place on which you stand is holy ground"* (Exodus 3:5). God is in a league alone, and although we are His beloved creation, we should always reverence and glorify Him.

Great is the Lord and greatly to be praised in the city of our God, in the mountain of His holiness. Beautiful for situation, the joy of the whole Earth is Mount Zion on the sides of the north, the city of the Great King. ... We have thought of Thy lovingkindness, O God, in the midst of Thy Temple. According to Thy Name, O God, so is Thy praise unto the ends of the Earth; Thy right hand is full of righteousness. ... For this God is our God forever and ever; He will be our Guide even unto death.

– Psalm 48:1-2, 9-10, 14; King James Version

Even our adversary the devil recognizes God's ultimate superiority: *"You believe that God is One. You do well; the demons also believe, and shudder"* (James 2:19).

People who humble themselves ***outside*** of the presence of God have fallen into the trap of Satan, whereby they lose all hope, all will to live. Look at the derelicts on Skid Row. But when you *"turn aside"* and humble yourself before Almighty God, that humble heart becomes the first component of His "Operation Rescue":

> *"Because your heart was tender and you humbled yourself before God ... and because you humbled yourself before Me, tore your clothes and wept before Me, I truly have heard you," declares the Lord.*
>
> – 2 Chronicles 34:27

> *The sacrifices of God are a broken spirit; a broken and a contrite heart, O God, Thou wilt not despise.*
>
> – Psalm 51:17

Key #3 – Understand What God Does

> *He said also, "I am the God of your father, the God of Abraham, the God of Isaac, and the God of Jacob." Then Moses hid his face, for he was afraid to look at God.*
>
> *And the Lord said, "I have surely seen the affliction of My people who are in Egypt, and have given heed to their cry because of their taskmasters, for I am aware of their sufferings. So I have come down to deliver them from the power of the Egyptians, and to bring them up from that land to a good and spacious land, to a land flowing milk and honey, to the place of the Canaanites, Hittites, Amorites, Perizzites, Hivites, and Jebusites.*
>
> *"And now, behold, the cry of the sons of Israel has come to Me; furthermore, I have seen the oppression with*

which the Egyptians are oppressing them. Therefore, come now, and I will send you to Pharaoh, so that you may bring My people, the sons of Israel, out of Egypt."

– Exodus 3:6-10

Realize that God has been orchestrating (writing and reading) the script of your life, and has even been "pulling the strings":

O Lord, You have searched me and known me; You know my downsitting and mine uprising, You understand my thought afar off. You have directed my path and my lying down, and are acquainted with all my ways. For there is not a word in my tongue, but lo, O Lord, You know it altogether. You have beset me behind and before, and laid Your hand upon me. Such knowledge is too wonderful for me; it is high, I cannot attain unto it.

– Psalm 139:1-6

But the very hairs of your head are all numbered.

– Matthew 10:30

Indeed, the very hairs of your head are all numbered. Do not fear; you are of more value than many sparrows.

– Luke 12:7

Who was it that allowed the jealousy in the hearts of Joseph's ten brothers > that caused them to sell him into slavery > that brought him to be sold to Potiphar in Egypt > that earned him the governorship of Egypt > that migrated the children of Israel to Egypt, in the first place? Luck? The "affairs of men"? Circumstance? Coincidence? "Only understandable in hindsight"?

Who was it that caused Moses to be born in the house of Levi > but entrusted to the adoption, care and upbringing of Pharaoh's daughter? Accident? Fate? Serendipity? Who awakened Moses' social conscience and allowed his abrupt actions to force him to flee from his "artificial environment" to the deserts of Midian? Moses perceived all these details as disaster, failure, loss. But ***God*** was orchestrating all of this to get Moses to the place where a humble shepherd could find the time to *"turn aside and see this strange sight"* (Exodus 3:3) ... and thus begin the road to the fulfillment of his true destiny!

It is *always* God's plan to bring us out of affliction into abundance:

> *So I have come down to deliver them from the power of the Egyptians, and to bring them up from that land to a good and spacious land, to a land flowing with milk and honey.*
>
> – Exodus 3:8

The only time we need to spend dangling at the end of that old hopeless rope is the amount of time it take for us to realize God is trying to get our attention!

Key # 4 – Know Who God *Really* Is

> *But Moses said to God, "Who am I that I should go to Pharaoh, and that I should bring the sons of Israel out of Egypt?"*
>
> *And He said, "Certainly I will be with you, and this shall be a sign to you that it is I Who have sent you: when you have brought the people out of Egypt, you shall worship God at this mountain."*
>
> *Then Moses said to God, "Behold, I am going to the sons of Israel, and I shall say to the, 'The God of your*

fathers as sent me to you.' Now they may say to me, 'What is His Name?' What shall I answer them?"

And God said to Moses, "I AM WHO I AM"; and He said, "Thus you shall say to the sons of Israel, 'I AM has sent me to you.'"

– Exodus 3:11-14

Here comes the part I personally like the best: **when God reveals Himself to us – *I Am Who I Am and What I Am, and I will be what I will be, because I Am"*** (Exodus 3:14).

There's an interesting passage in Matthew 4 where Satan attempts to duplicate God, where he tries to imitate the Almighty to impress and persuade Jesus. The Deceiver focuses on **things:** *"And the tempter came and said to Him, 'If You are the Son of God, command that these stones become bread'"* (Matthew 4:3) ... then on **circumstances:** *"Satan said to Him, 'If You are the Son of God, throw Yourself down, for it is written, 'He will give His angels charge concerning You'; and on their hands they will bear You up, lest You strike Your foot against a stone"* (verse 6) ... and finally on **reputation:** *"Again, the devil took Him to a very high mountain, and showed Him all the kingdoms of the world, and their glory; and he said to Him, 'All these things will I give You, if You fall down and worship me'"* (verses 8 to 9).

But God did ***not*** say: *"I will dazzle You with My ability to provide for your needs -- although it is true I 'own the cattle on a thousand hills'* (Psalm 50:10) *... or I will amaze you with My ability to supernaturally change your circumstances* (see Daniel 3:22-25; Jonah 1:17; Acts 16:25) *... or I will exalt you before all men* (see Job 36:7; 2 Samuel 2:8; 1 Kings 5:5; 2 Chronicles 23:20; Psalm 132:11;

Jeremiah 52:32; Revelation 3:21)." God, the Almighty Creator, identifies Himself simply as Himself – He doesn't need a pedigree or list of accomplishments.

This also relates to what is essentially the difference between our offerings of thanksgiving and praise – which are for what He has done for us – and our offerings of worship – which is adoring Him for Who He is.

We Seek Your Face

(verse 1) *Not power or glory, not ministry or fame*
Not ableness or eloquence, not titles or name
Not fire or miracles, not thunder or rain
We seek Your face, we seek Your face, we seek Your face

(verse 2) *Not crowns or kingdoms, not houses or land*
Not passions or pleasure, not blessings from Your hand
Not earthly inheritance or richest reward
We seek Your face, we seek Your face, we seek Your face

(chorus) *All that we are and ever hope to be*
We lay it at Your feet in worship
Poured out to You as a living sacrifice
In worship, worship to You
We seek Your face, we seek Your face, we seek Your face

(finale) *Not power or glory, seek Your face*
Not crowns or kingdoms, seek Your face
Not passions, pleasures, seek Your face
We seek Your presence, seek Your face

– Parachute Band, 1998

Now, having come humbly into the presence of God, He can begin His work of reconstruction, showing us how to shimmy up that hopeless rope and find our footing again.

That's When

Let's focus on the main purpose of this book – ***"What's that YOU have in your hands?"*** – and to do so, I would like to introduce this theme with the words of a contemporary song:

(Verse 1) *When you have a work to do*
And the task ahead seems bigger than you
That's when He steps in
When you know in your heart that God's command
Takes more than can be done by Man
That's when He steps in
He sees you at the point of your need
He sees you at the point of crossing your Red Sea
In the moment you call, when you've given you all
He steps in, He steps in
And He'll say

(Chorus) *"What's that you have in your hands?*
I can use it, if you're willing to lose it
Take the little you have and make it grand
I am El-Shaddai
And I'll more than supply your need"

(Verse 2) *When all you have is oil in a jar*
That's a reflection of where you are
That's when He steps in
A little boy's lunch of fish and bread
Is all you have for the need ahead
That's when He steps in
Let Him take it and bless it and break it and give it
He'll multiply it in the moment you live it
And in the moment you call, when you've given your all

He steps in, He steps in
And He'll say

(Chorus) *"What's that you have in your hand?*
I can use it, if you're willing to lose it
Take the little you have and make it grand
I am El-Shaddai, and I'll more than supply your need
I will supply your need!"

– Helena Barrington

So what's that you have in ***your*** hands?

Chapter 3: In Celebration of Hands

I want to segue for just a moment, and write about hands.

Bob Gossett's Hands

In 1937, when Bob Gossett was a young man in Depression-era Oklahoma, he was fortunate to have work as an itinerant electrician. Although it often meant traveling away from town to town like a nomad, it also provided a passable living for his wife and three little children. When he worked outdoors, climbing the spikes of electrical poles up twenty and thirty feet in the air, a wide leather safety belt hung with the tools of his trade around his waist, Bob knew he was doing good work.

One day, Bob was perched high above the ground repairing a transformer, when his grounds-man hoisted up the next electrical cable in the series to be strung ... unintentionally handing Bob a "live wire." Bob innocently took the wire in his right hand – and 2,400 volts of electricity shot through his body, instantly igniting his right hand and arm, and burning the safety belt in two. Shocked into unconsciousness, Bob fell down the thirty-foot drop, landing face-down on a pile of rocks at the base of the pole.

Co-workers raced to his inert body but were momentarily repelled by the stench of burned human flesh: Bob's skin had turned completely black from head to toe. One buddy watched as the fingers of Bob's right hand crumbled into ashes. Someone reached into Bob's mouth and pulled out his swallowed tongue, and in their rudimentary way, they performed basic life-saving measures. They bundled him into the "rumble seat" of a Model-T car, and rushed him to the hospital at Cushing,

Oklahoma.

Bob's life hung in the balance for weeks. Doctors amputated the remnants of his right arm just below the elbow. His left arm was spared amputation, but remained without sensation for the rest of his life. After many months of therapy, now fitted with a prosthetic arm, Bob returned to his little family, his career as an electrician over.

Bob's son Don was eight years old when this tragic accident occurred. Mostly glad his father was not dead, Don grew up with automatically helping his father tie his shoes or button his shirt – simply compensating with his own two good hands for his father's missing right hand and the damaged, clumsy left hand.

It's no wonder, then, as Don Gossett grew into manhood, he truly appreciated the ability – as a young athlete – to accurately shoot a basketball from his two good hands ... or – as the author of many Christian books – to place his two good hands on the keys of a typewriter, to write a manuscript ... or – as a minister of the Gospel of Jesus Christ – to lay his two good hands on the peoples of the world in obedience to *"the ministry of laying on of hands"* (see Mark 16:18).

> *Withhold not good from them to whom it is due, when it is in the power of thine hand to do it.*
>
> – Proverbs 3:27; KJV

Bob Gossett was my paternal grandfather. I was never afraid of his prosthetic hand.

Michelangelo's Painting

In my opinion, Michelangelo di Lodovico Buonarroti

Simoni (1475 - 1564) was the most talented European artist of the 15th century. His sculptures, paintings, poetry, and architectural designs remain some of the world's most exquisite art treasures. One of my personal favorites is his painting of *"The Creation of Adam"* on the ceiling of the Sistine Chapel in Rome. Just a portion of the masterpiece that encompasses every theme from "the creation of the world" to "the last judgement," has this unique story behind it.

When Michelangelo was a young apprentice, he was trained in various forms of sculpture, which included hands-on work in the quarries, where great blocks of pure-white marble were cut from the mountain. A life-long family friend was Topolino of Settignano, the master stone-cutter, who worked closely with the young artist to teach him how to find the most perfect stones and safely bring them down the mountainside – a difficult, dangerous occupation. Michelangelo's own father was a cold-hearted, distant man, and young Michelangelo often turned to Topolino and his family for solace, friendship and nurturing.

Years later, when Michelangelo was given the commission to paint the ceiling of the Sistine Chapel, he remembered the loving, paternal care of his friend Topolino – and recalled his massive, strong, stone-cutter's hands. From memory, Michelangelo painted Topolino's likeness as Jehovah God ... and Topolino's sinewy, sure-fingered, strong right hand as the hand of the Creator, reaching out across eternity with one firm forefinger to touch the outstretched finger of the sleeping, recumbent Adam.

The hands of Adam? Michelangelo – perhaps the world's greatest artist – looked at his own hands and painted them for Adam, receiving the gift of life from God the Father.

Joyce's Baby

During her third pregnancy, Joyce developed a strange condition whereby her body absorbed too much calcium while her baby absorbed little or no calcium – a condition which was not readily identified until after the birth of the child. The birth itself was not an easy process. In the small town where Joyce and her husband lived, there were not too many doctors. As her delivery date arrived, their regular family doctor was out-of-town at a medical conference, so another doctor-on-call at the small hospital was assigned to Joyce.

However, the doctor-on-call also had a secret problem with alcohol. As Joyce quickly progressed through labor and birth was imminent, the nurses – who technically were forbidden to deliver babies – scrambled frantically to locate the doctor, only to eventually find him passed out drunk. Even as the baby's head was crowning, nurses placed pads and pillows between Joyce's legs, preventing delivery.

Finally, another doctor was found and the baby was born: a second daughter for Joyce and her husband. But this child was not healthy and strong like the first two babies: she had "wet lungs" (most likely from aspirating birthing fluids during the delayed delivery) and a blue complexion ... and the soft, calcium-deprived bones of her limbs resulted in clubbed feet (turned perpendicular at the ankles) and deformed hands (bent backward at the wrist, resting against the forearms), also contributed by reverse pressure forced during the prevention of delivery.

The infant was immediately placed in an incubator to assist her difficult breathing, and quickly her tiny feet were confined in plaster casts in an effort to straighten the still-soft bones. Nurses were assigned to therapeutically massage the hands and wrists of the baby, trying to coax her hands back into a normal condition.

Joyce's health deteriorated after the birth as the excess calcium settled in her joints, inflaming them; she developed a condition known as "rheumatic fever." Eventually she was discharged from the hospital, sent home to total bed-rest for many months. That left her husband with the care of a two-year-old boy, a one-year-old girl ... and a hospitalized newborn daughter. Daily, Joyce's husband visited the sickly infant, pushing his big hands through the portals of the incubator to caress her trembling body, to touch and hold her as much as possible ... and to lay on his hands in fervent, believing prayer.

After six weeks, the baby girl was released from the hospital. Some weeks later, Joyce and her husband were told, "You need to bring the baby back so we can change the casts on her feet for a larger size. All babies' feet grow quickly." When the miniature plaster casts were cut apart – accidentally slicing the outer skin, resulting in lifelong scars – the doctors were amazed to discover the clubbed feet condition was gone, the feet were unexpectedly normal. The doctors were pleasantly surprised ... but the parents strongly believed God has healed their daughter supernaturally.

And the baby's hands were normal too. In time, the little girl grew up and learned to use her two good hands to play the piano, to hold a pen and write or to type quickly on a keyboard ... and eventually, to hold and caress her own little children, and later her grandchildren.

I am that little girl: Jeanne Michele Gossett (now Halsey). My parents, Don and Joyce Gossett, always called me “Jeanne Michele, Miracle Girl.” I was healed by the supernatural intervention of God!

In 1987, at Wally and Marilyn Hickey’s “Happy Church” in Denver, Colorado, I was asked by visiting evangelist Benny Hinn (a longtime Gossett family friend) to assist him in the prayer line. Suddenly, at one point during the healing ministry, Benny whirled around and grabbed both my hands in his and began to pray over them. Immediately my hands felt red-hot as Benny prophesied, “These hands shall write books that will touch the lives of millions of people around the world, pointing them to Jesus Christ!” My hands were made for the service of God ... and, as a writer, I **have** written many books.

Hands At An Airport

While writing this book, I spent a lot of time in airports. As a writer, I have always been an avid “people-watcher,” but during this period, I found myself unintentionally watching hands.

Traveling people always have things in their hands: tickets and boarding passes; the inevitable carry-on luggage that comes in all shapes, sizes and styles; walking canes or umbrellas; modern-day nomads carrying bottled water, paperback novels, newspapers, reading materials, to pass away the hours.

Then there are parents firmly holding the hands of their little children, willing the wide-eyed kids to stay close despite the high-tech circus around them. There are other hands eagerly waving “hello,” reaching out to welcome and embrace ... and subdued gestures as they wave “goodbye,” where hands often fall forlornly to the side after.

I watched an elderly couple, waiting. The were very well-dressed and obviously financially secure, but he was sitting in a wheelchair and she was carrying a small case with some sort of necessary medical apparatus. "A long life together," was my thought, "lived as husband and wife, partners, friends."

It was their elderly hands that most fascinated me. Youthfulness was long gone from their hands: no amount of skin care, nail care, massage therapy, designer clothes, not even elegant gloves or expensive jewelry could disguise the agedness of their hands. Perhaps his were the hands that signed multi-million dollar industrial contracts; perhaps hers were the hands that flashed large-carat diamonds at high-society functions.

Now they both had hands gnarled with age, heavily veined, quavering constantly. His hands rested on the arms of the wheelchair, fingers sometimes twitching as we waited for our flights to be called. Her hands fussed with the locks, zippers and straps of the small case, as though seeking reassurance that the medical contents were safely within.

Then their hands brushed and met between them ... and they gently clasped them together. Not looking at one another, the gesture was so familiar, and they sat together calmly, their years of bonding uniting them in quiet peace. They comforted each other with the touch of their hands, wordless communication of "two halves of one whole."

I marvel how God has shaped human hands to fit together so perfectly.

Chapter 3: ***"What's That You Have In Your Hands?"***

Now back to our theme: the recurrent pattern how God helps us pull up from the end of that hopeless rope and walk into His provision and victory. This is universally called "the law of sowing and reaping." I'm not really addressing finances at this point – although Christians are much-taught about how financial blessings result from faithful tithing and generous giving – but more about stepping past finances to find the gifts God has already placed within us. These gifts are within our own hands, and need to be unlocked, activated, utilized, and benefitted from. I'll cite several Biblical examples to illustrate this principle ... but first, let's return to Moses.

Moses, Again

Let's start by reviewing Exodus 4:

> *And the Lord said to Moses, "What is that in your hand?"*
>
> *And he said, "A staff."*
>
> *Then He said, "Throw it on the ground." So he threw it on the ground, and it became a serpent; and Moses fled from it. But the Lord said to Moses, "Stretch out your hands and grasp it by its tail" – so he stretched out his hand and caught it, and it became a staff in his hand – "that they may believe that the Lord, the God of their fathers, the God of Abraham, the God of Isaac, and the God of Jacob, has appeared to you."*
>
> – Exodus 4:2-5

Remember how discouraged and hopeless Moses was when he wandered into the wilderness of Horeb? He had lost everything: his family (Pharaoh and Company), his fortune (he was no longer the prince of Egypt), his reputation (he was now an accused murderer and a fugitive). He was an outcast, reduced to the mindless job of tending the sheep of his father-in-law Jethro, the (pagan) priest of Midian. He was not even among his own people, the Hebrews, who believed in the One God of Israel. And his people were suffering greatly as the enslaved race who served the polytheistic Egyptians (see Exodus 3:23-25) ... and Moses felt a certain liability for the sorry state of the Hebrews.

Moses, clutching his stick, shuffling along after the flock of sheep that didn't even belong to him, came around to the west side of the mountain ... and there he met God. God was waiting for him – *"God called to him from the midst of the bush, 'Moses, Moses!'"* – then identified Himself – *"'I am the God of your fathers'"* – and then Moses malfunctioned – *"Then Moses hid his face, for he was afraid to look at God"* (Exodus 3:4-6).

A Modern Problem

Why do so many people today discount the supernatural? Who do many ignore that "God-shaped place" (thank you, Blaise Pascal) in their hearts, claiming agnosticism or atheism or "religion is for the feeble-minded"? I think these people are, like Moses, *"afraid to look at God"* because He is too holy, too wonderful, too far beyond their scope of understanding. *"My ways are far above your ways"* (Isaiah 55:8-9). They cannot fathom His supernatural nature, His purity and righteousness ... and they don't want to admit to their own shortcomings, their inherent unrighteousness:

This I say therefore and testify in the Lord, that ye henceforth walk not as other Gentiles walk, in the vanity of their mind, having the understanding darkened, being alienated from the life of God through the ignorance that is in them, because of the blindness of their heart.
– Ephesians 4:17-18

They pretend they don't believe in Him so they won't have the face the issue of their own sin-nature. "What you don't know can't hurt you" – that's one of Satan's tricks of ignorance that regularly deceives people. But I know **all** men are spirit-beings living within a flesh-shell, and absolutely everyone is created with that place in their spirit which connects directly with God, Who is Spirit (John 4:24). It is so easy to spend a lifetime stuffing "plugs" – intellectualism, materialism, even religion – into that opening, damming the natural flow meant to be between God and His creation, Man.

FIRE!

When God wanted to finally catch Moses' attention, He started a fire: *"And the angel of the Lord appeared to him in a blazing fire from the midst of the bush; and he looked, and behold, the bush was burning with fire, yet the bush was not consumed"* (Exodus 3:2). When He had his attention riveted, when He called out to the heart of the man (see verse 4), when He placed him in a position of humility (see verse 5), then He began to talk with him about the future.

Then Moses began to ask all the **wrong** questions! God had just told him Who He was and what his relationship to Him was, and what plans for his life He had ... and Moses shortsightedly asked the usual selfish question, “Who, me?”

Is our flesh to tender and precious to us that we regard it more valued than someone else’s? Are the circumstances of our lives so important, so demanding, so heavy, that we cannot lift up our vision beyond ourselves? One of the most difficult concepts to grasp and implement is learning to see with Heaven’s view.

Through Heaven’s Eyes

A single thread in a tapestry, though its color brightly shine
Can never see its purpose in the pattern of the grand design
And the stone that sits on the very top of the mountain’s mighty face
Does it think it’s more important than the stones that form the base
So how can you see what your life is worth or where your value lies
You can never see through the eyes of Man
You must look at your life, look at your life through Heaven’s eyes

A lake of gold in the desert sand is less than a cool fresh breeze
And to one lost sheep, a shepherd boy is greater than the richest king
If a man lose everything he owns has he truly lost his worth
Or is it the beginning of a new and brighter birth
So how do you measure the worth of a man, in wealth or strength or size
In how much he gained or how much he gave
The answer will come, the answer will come to him who tries
To look at his life through Heaven’s eyes

And that's why we share all we have with you, though there's little to be found
When all you have is nothing, there's a lot to go around
No life can escape being blown about by the winds of change and chance
And though you may never know all the steps, you must learn to join the dance
You must learn to join the dance

So how to you judge what a man is worth, by what he builds or buys
You can never see with your eyes on Earth, look through Heaven's eyes
Look at your life, look at your life
Look at your life through Heaven's eyes.

– Stephen Schwarz and Hans Zimmer

For the remainder of Exodus chapter 3, God patiently explains to Moses what was His plan of salvation (see verses 19 to 20) ... but in Exodus chapter 4, Moses continued to argue and complain and dispute with God (see verses 1 through 13). Doubt, unbelief, self-importance (which goes hand-in-hand with self-inadequacy) always interfere with our ability to hear God, to obey His Word. It is so typical to be so consumed with our **selves** – our lives, our problems – that we prevent ourselves from connecting with God.

"Hindered Prayers": A Light Comedy Pantomime

Theme: Contrasts between FAITH and DOUBT.
Foundation: James 5:16; 1 Peter 3:7, 5:7.

Cast: NARRATOR (male or female) ... GOD ... DOUBT-

FULL PRAY-ER (male or female; although presented here as female) ... FAITH-FULL PRAY-ER (male or female; although presented here as female).

Director's Notes: The "sack of problems" which each PRAY-ER carries doesn't have to be an elaborate prop; in fact, I suggest using a typical woman's purse (with lipstick, candy mints, wallet, checkbook, car keys, crumpled-up tissue, etc.) or a backpack. The trick is to make it seem that the "load" weighs about a ton, and that each of these mundane things of life represent extremely huge problems. (Author's Note: When performing this skit at Women's Conferences, I asked for the use of the purses of two unsuspecting women in the audience, and then handed them – without any prior knowledge of their contents – to the actors; it made for some slightly embarrassing humor, which most women actually enjoyed!)

Narration and Pantomime Action

NARRATOR: One of the most common complaints Christians have today is that "God doesn't hear them when they pray." The problem does not lie with God's ability to hear at all – it has to do with our approach to God. For instance, have you ever personally experienced something like this scenario?

PANTOMIME ACTION (this can be modified for narration also, if necessary): GOD is seated high on His Throne. Enter DOUBT-FULL PRAY-ER, heavily burdened by a sack of "problems," nearly bent over, unable to see GOD's face. DOUBT creakily kneels at GOD's feet, painfully unloads and pours out her sack of "stuff." Instead of first worshipping GOD, confidently and trustingly address Him and *then* offering her problems to Him, she anxiously begins to place each bit and piece into an imaginary large cauldron. When everything is in the pot, DOUBT uses a large spoon or ladle to stir her problems all around, sometimes taking out pieces to examine and "complain" about them. All the while, GOD is watching this patiently but with a touch of dismay, unable to intervene because DOUBT is too busy complaining.

Eventually, DOUBT begins putting back the pieces into her sack, stiffly resumes the same heavy load on her back, and tiredly shuffles away. Nothing has changed; her prayers are unsatisfied.

NARRATOR: Notice that DOUBT-FULL PRAY-ER never actually gets face-to-ace with GOD since she – because of her doubt – just can't see past her problems. For some reason, she has forgotten that GOD is the Creator of the entire Universe ... and there is nothing too big or too difficult or too unsolvable for Him, and certainly there is nothing that ever surprises Him. Instead, her focus remains on her problems and her doubts, and she misses out on that necessary, intimate contact – the joy of worshipping Him – with the One Who loves her, Who knows her, Who wants what is best for her at all times ... and Who knows how to solve her problems!

Is there an easy answer to this scenario? Not exactly, because everyone's situations are different. But there is one outstanding basic principle that **always** applies to our prayers: the importance of beginning with faith, trust and worship. Like this ...

PANTOMIME ACTION: Enter FAITH-FULL PRAY-ER, also heavily burdened with a sack of problems. But when FAITH kneels at GOD's feet, she does whatever it takes to peer up past her load, seeking eye contact. Then GOD lovingly takes the sack off her back and sets it aside. FAITH is now able to stand up straight and tall, looking directly, clearly, joyfully into GOD's face. For a long while, GOD and FAITH simply gaze at each other with mutual love and joy; occasionally GOD strokes FAITH's face or pats her shoulders, ministering strength and encouragement.

After awhile, GOD picks up FAITH's sack and pours out all the stuff onto the table. Then together, GOD and FAITH pick up various bits and pieces, arranging them like a chessboard, examining them together, "discussing" ways to deal with the problems; sometimes discarding "trash" pieces.

Eventually FAITH, looking peaceful and restored, starts putting back some (not all) the pieces into her sack. Instead of shouldering her burden with the same painful, bent-over method, now she simply swings the sack lightly at her side, cheerfully waving farewell to GOD as she departs.

NARRATOR: God, in actuality, is a jealous God – He doesn't want **anything** else to come before Him. But when we allow our cares, difficulties and troubles, our doubts and lack of faith, to overwhelm our ability to ***see Him first*** ... then we are making those problems greater than He is in our perception; we are allowing them to take first place in our lives ... and that's a no-no!

That's why He instructs us to *"Cast our cares upon Him, for He cares for us."* Because He is **passionate** about fellowship with His children, and our worship of Him is essential to a healthy relationship with Him, He has designed the means whereby we can live on this sin-filled Earth – with all its problems, trials, disappointments, hurts, pains, sorrows, tragedies, disasters ... well, you get the picture! – as genuine Christian people filled with faith. We do not escape our trouble-filled lives – that is character development, folks – but we don't allow those troubles to overwhelm us and block the free flow of fellowship with Him ... not ever!

So the next time you are heavily burdened with your load of life, and you need to talk it over with your Heavenly Father, remember your priorities: God comes first. He is worthy of our trust, He responds to our faith. **Begin** with humble worship, for He **does** have the answers for your life.

God's Proof

God had promised Moses there would be miracles in Egypt – *"So I will stretch out My hand, and strike Egypt with all My miracle which I shall do in the midst of it"* (Exodus 3:20) – which would enable him to accomplish what He sent him to do. But Moses didn't get it; he didn't believe it, or he needed proof – *"Then Moses answered and said, 'What if they will not believe me, or listen to what I say? For they may say, 'The Lord has not appeared to you'"* (4:1). So God gave him proof:

And the Lord furthermore said to him, "Now put your hand into your bosom." So he put his hands into his bosom, and when he took it out, behold, his hand was leprous like snow.

Then He said, "Put your hand into your bosom again." So he put his hand into his bosom again, and when he took it out of his bosom, behold, it was restored like the rest of his flesh.

"And it shall come about that if they will not believe you or heed the witness of the first sign, they may believe the witness of the last sign. But it shall be that if they

will not believe even these two signs or heed what you say, then you take some water from the Nile and pour it on the dry ground, and the water which you take from the Nile will become blood on the dry ground."

– Exodus 4:2-9

***"What is that in your hand?"* asked God.**

Moses looked all around in bewilderment. He no longer wore the signet ring of the son of Pharaoh, which had been his royal mark of identity and authority. Just the flash of that signet ring would buy him the fastest horse in the Royal Stable ... or the command of the swiftest boat in the Royal Navy ... or cause a troop of the finest soldiers of the Royal Army to fall into step behind his Royal Chariot.

Although the Bible only records it once (after all, this is Moses writing about himself, many years later!), I sometimes wonder just how many times God had to ask that specific question of Moses. Obtuseness is an all-too-human trait. How long do you think it took for Moses to reach that point of being able to say – as the apostle Paul would state so many years later – *"I can do all things through Christ Who strengthens me"* (see Philippians 4:13)?

So God asked, *"What's that in your hand?"* ... and Moses dully answered, *"Just a lifeless old shepherd's staff."* **Then** God commanded, *"Cast it down!"*

The Principles of "Sowing Out of Your Need"

Throw it away! Give it away! A curious principle of life is: *"Whatever need or lack you have in your life own, find someone else with that same lack or need ... and minister to them."* Do you have an illness, such as cancer?

Find someone with cancer and go pray for them. Do you have a financial shortfall? Find someone monetarily challenged and give them something, even if it just a small amount. Do you have a relationship problem? Find someone with a similar challenge and just listen to them, encourage them, empathize with them. These are the seeds you plant for the harvest you need in your own life!

"What do you mean, Jeanne?" There was a time when I was experiencing pain in my joints, especially my knees. One Sunday, toward the end of church, several people came to the front of the church for prayer for various reasons. For awhile, I simply sat in my seat – unconsciously rubbing my aching knees – and praying intercessively for those being ministered to at the front. Then I felt compelled to leave my seat and pray with a woman kneeling at the altar. I knelt down beside her and poured my whole being into praying for this lady. When I arose from praying, I realized there was no pain in my knees at all! And the pain did not come back.

I was not focusing on my problems in a typically selfish manner. I chose to make myself available to God, to "become His hands made flesh" in ministry to others. While I was "distracted" by ministering to another, God took the opportunity to heal me of my own problem ... and I had not even asked Him!

Although I don't have a whole lot of Scriptures to back up this principle, I believe it demonstrates Christ-like compassion for others, revealing God's sense of humor to touch us when we're "not looking," indicating in us a spirit of generosity that pleases Him:

The generous man will be prospered. And he who waters will himself be watered.

– Proverbs 11:25

An Illustration of Faith

Another way to look at this principle is that God was asking Moses to "throw away" his last defense ... to let go of the end of the hopeless rope. Free-falling at the command of God – also known as "walking by faith" – is an often-repeated exercise in the life of the maturing Christian.

One of my favorite modern-day illustrations of "faith" is depicted in the popular movie, *"Indiana Jones and the Last Crusade."* In the 1989 film, our fearless hero – archeologist/adventurer "Dr. Henry 'Indiana' Jones, Jr." – goes through a series of escapades and difficult situations while trying to locate the famed "Holy Grail." In a scene of immense tension, while under a strong time-constraint – his own father's life is at risk, at this point – "Indy" is confronted with a wide, dangerous, bottomless chasm which he must cross, although there is no apparent way to do so.

He believes, however, in the instructions assembled by his father's research. Indy stands for a moment, hovering on the edge of the deep chasm that plunges between two sheer cliff-faces. Then he slowly extends one booted foot out into thin air, holds it steady for another moment, then brings that foot down onto ... "nothing" – only to discover a narrow causeway straddling the deep gorge, which he could not previously see because it was **disguised** to look exactly like the face of the opposite cliff!

That is faith: when you are willing to step out into thin air for what you believe in – even if you cannot **see** it – only to find God has **already** prepared the way for you to cross the challenges of your life! Like Indy's erudite father, our Heavenly Father has put together a detailed written instruction-book to help us live – we only have to believe Him and trust His Word! *“Blessed are those who have not seen, and yet believed”* (John 20:29).

Discover what you have in your hands ... and give it to God to use miraculously.

Chapter 4: **Elijah and the Widow**

This Bible story is found in 1 Kings 17:1-16, and is very familiar. Here I will tell it in my own way; being a theatrical sort of person, I will recount it somewhat dramatically as well.

First, the Word:

Now Elijah the Tishbite, who was of the settlers of Gilead, said to Ahab, "As the Lord, the God of Israel lives, before Whom I stand, surely there shall be neither dew nor rain these years, except by my word."

And the word of the Lord came to him, saying, "Go away from here and turn eastward, and hide yourself by the Brook of Cherith, which is east of the Jordan. And it shall be that you shall drink of the brook, and I have commanded the ravens to provide for you there."

So he went and did according to the word of the Lord, for he went and lived by the Brook Cherith, which is east of Jordan. And the ravens brought him bread and meat in the morning, and bread and meat in the evening, and he would would drink from the brook.

And it happened after a while, that the brook dried up, because there was no rain in the land. Then the word of the Lord came to him, saying, "Arise, go to Zarephath, which belongs to Sidon, and stay there; behold, I have commanded a widow there to provide for you."

So he arose and went to Zarephath, and when he came to the gate of the city, behold, a widow was there gathering sticks; and he called to her and said, "Please get me a little water in a jar, that I may drink." And as she was going to get it, he called to her and said, "Please bring me a piece of bread in your hand."

But she said, "As the Lord your God lives, I have no bread, only a handful of flour in the bowl and a little oil in the jar; and behold, I am gathering a few sticks that I may go in and prepare for me and my son, that we may eat it and die."

Then Elijah said to her, "Do not fear; go and do as you have said, but make me a little bread cake from it first, and bring it out to me, and afterward you may make one for yourself and for your son. For thus says the Lord God of Israel, 'The bowl of flour shall not be exhausted, nor shall the jar of oil be empty, until the day that the Lord sends rain on the face of the Earth.'"

So she went and did according to the word of Elijah, and she and he and her household ate for many days. The bowl of flour was not exhausted nor did the jar of oil become empty, according to the word of the Lord which He spoke through Elijah.

– 1 Kings 17:1-16

Flour and Oil – Scene One

Cast: the LORD GOD ... ELIJAH, the Prophet ... AHAB, the king of Israel (not seen) ... the GENERAL POPULATION of Israel (not seen), scraping through a terrible drought in the land ... the WIDOW of Zarepheth ... the Widow's YOUNG SON (not seen) ... the Widow's HOUSEHOLD (not specified, but likely the farm-hands who had no where else to go in the drought)

Our story begins with the LORD GOD speaking to ELIJAH, now hungry and thirsty after a long season of drought.

GOD (off-stage voice only): Elijah, My servant, I have a plan for you.

ELIJAH (croaky, dry voice): Yes, Lord?

GOD: I know you are hungry, and since the Brook of Cherith has now dried up, I know you are thirsty too.

ELIJAH: You've got that right, Lord!

GOD: So I have already gone before you and made a provision for you ... at Zarephath.

ELIJAH: Mighty glad to hear that, Lord.

GOD: Although she doesn't know it yet, I have arranged for a little Widow woman to take care of your needs.

ELIJAH: Excuse me? *"She doesn't know it yet?"* (Pause; he's thinking) She must have plenty of reserves, right?

GOD (patiently): Not exactly. In fact, she too is at the bottom of her barrel, so to speak, and has given up hope, and is preparing to die.

ELIJAH (sighing after a moment's pause): You're planning another one of those supernatural things, aren't You?

GOD (chuckling): You betcha!

ELIJAH (as he gets up, gathers his things and prepares to leave for Zarephath): Okay, Lord, anything You say!

I've heard that the prayer-life of some of the greatest servants of God consists less of talking and more of listening. The longer we are in relationship with God, the more we know His personality and character. Those long periods of companionable, speechless interludes become refreshing, nourishing, pleasurable, and so instructive.

There was another aspect of God's personality that Elijah had learned to rely upon: ***expect the unexpected.*** When God said, ***"I have already gone before you*** *to prepare"* – that is another of His distinctive traits!

FLOUR AND OIL – SCENE TWO

Just outside the Gate of Zarephath. The WIDOW, a tired-looking, skinny woman wearing dusty, once-fine garments, is slowly gathering a bundle of sticks. Enter ELIJH, travel-worn, sweaty and weary.

ELIJAH (to the WIDOW): Excuse me, ma'am, but is this the City of Zarephath?

WIDOW (slowly turning around; deeply mournful, almost trance-like): Yes, these are the Gates of Zarephath ... or what is left of our once-proud city. (She sighs and stoops back to her wood-gathering)

ELIJAH looks up to Heaven and briefly pantomimes a conversation with GOD, pointing at the WIDOW, then shrugging and proceeding with the plan.

ELIJAH: I am sorry to bother you again, ma'am, but could I trouble you for a little drink of water? I have traveled here all the way from the Brook of Cherith and ...

WIDOW (peering at ELIJAH more closely; then she "remembers her manners" – she was once a very refined lady): Of course, stranger. I will be just a moment. (She turns to leave)

ELIJAH (encouraged): And please bring me a tiny piece of bread in your hand.

WIDOW (suddenly angry; she drops her bundle of sticks and puts her fists on her hips in exasperation): Aw, c'mon, mister! As the Lord God knows ... (ELIJAH surreptitiously sneaks a tiny wink Heaven-ward) ... I have no bread! Nobody has any bread! Nobody has any food at all! This drought that was decreed by that stupid, careless, wild-man prophet guy – I forget his name – has just brought everyone to complete ruin!

ELIJAH (somewhat alarmed at this outburst): What I meant was ...

WIDOW (ignoring his humble apology): I was once one of the wealthiest land-owners in Zarephath, but now that is all gone. My fields of barley are nothing but stubble, dead, dry and gone. My field-hands and servants have all tried to find work elsewhere, anywhere – just trying to find food! We have lost everything ... and there is nothing left to live for. My son and I, we are going to die. Period.

ELIJAH (now very alarmed): But ... but ...!

WIDOW (still indignant): And I mean we are going to die, literally! Do you know what I was doing out here when you showed up and asked for water and bread? No? I was gathering sticks so I could make a small fire. Then I was going to use the very last scrapings of flour in the bin and the very last droplets of oil in the jar to make the very last cake of bread for my son and me. Then I was going to bake the bread over this pitiful little fire. Then we were going to eat that lat bit of bread and drink the last drops of water, and then we were going to lay down and wait for death! (She runs out of steam)

ELIJAH: But I have been sent ...

WIDOW (continuing in a low voice): But then you come along and ask for our very last piece of bread ... (she suddenly flares again) ... how dare you ask me for that!

ELIJAH (taking control of the situation; slowly and distinctly): I have come to you at the command of the God of Israel. (The WIDOW is struck dumb) And I have come to say to you: *"Do not fear!"*

WIDOW (now suspicious): Are you ... are you that prophet-guy?

ELIJAH (crisply): Yes, I am, and I say to you: *"Go, and do as you have said."*

WIDOW (puzzled): Huh?

ELIJAH (repeating): *"Go, and do as you have said."* But make me a little bread cake first and bring it out here to me. Afterward, you may make one for yourself and for your son.

WIDOW (as if speaking to a child): You don't get it. There is not enough flour and oil to make one decent bread cake, and certainly not enough for two ...

ELIJAH (in a powerful voice): *For thus says the Lord God*

of Israel, "The bowl of flour shall not be exhausted nor shall the jar of oil be empty, until the day that the Lord sends rain on the face of the Earth."

For a long moment, the WIDOW just stares at ELIJAH, obviously processing what he said. Then a look of determination crosses her face and she turns to hurry through the city gates. ELIJAH just stands there, looking uncomfortable -- *"Maybe she was the wrong widow?"* -- when she abruptly reappears.

WIDOW (picking up her dropped sticks): Forgot the wood for the fire. I'll be right back. (She hurries away)

ELIJAH (biting his lip): Was that alright, Lord?

The WIDOW abruptly returns again, a cup of water in her hand.

WIDOW: Here's your water. (She starts to turn away again, then reconsiders) Hmm ... would you like to come to my house and wait for your bread there?

ELIJAH (with a big, relieved smile): I would be honored!

The End

We all know this story. The flour never ran out and the oil was constantly renewed. There was enough to sustain the Widow *and* her son *and* the Prophet *and* the rest of her household: *"So she went and did according to the word of Elijah, and she and he and her household ate for many days. The bowl of flour was not exhausted nor did the jar of oil become empty, according to the Word of the Lord which He spoke through Elijah"* (1 Kings 17:15-16) – for the duration of the drought.

So what did the Widow have in her hands? Well, at first she had firewood. You see, this woman was resigned to dying, but she wanted to go out in style. She was going to cook her bread – not just consume the last of her flour and the last of the oil uncooked. Maybe because she loved her son and still desired to do her very best by him. Maybe she half-though the cooking fire, if left unattended in that dry, drought-stricken climate, might burn down the whole house, thus disposing of their bodies dead already from starvation.

I think the face the Widow was out gathering firewood was significant. It is quite possible there was more than one beleaguered widow in Zarephath, and the sign Elijah needed to identify which widow was God's chosen vessel was, in act, her attitude of activity. Perhaps all the other widows were sitting passively, hopelessly, in their houses waiting to die. Perhaps too there are some passive Christians just waiting for God to solve their problems.

"Beggar-Believers"

There is a common misconception that *"God helps those who help themselves"* is a Scripture verse. It is not – but like many basic pearls of human wisdom, it is based on Godly wisdom. Far too many believers are lazy, even dishonest, in their approach and expectation of God's provision. They are beggars in the Kingdom of God, but they should not be.

I imagine every pastor can relate to the following true story (which I personally witnessed). After a messy divorce, a young woman with two young children went from church to church requesting financial aid. She would start at one church, looking sweet and much-wronged, and volunteer to teach Sunday School, or play the piano for the choir – but in exchange for regular monthly support.

After awhile, she became too "stressed out" as a single parent to continue her helpful contributions to the life of the church, but she still expected her stipend, even complaining to others about how meagerly it was apportioned to her. When church leadership suggested she secure some part-time job to help augment her income and improve her circumstances, she protested, claiming fragile health and quoting the verses about *"caring for widows and orphans"* (see 1 Timothy 5:3).

Eventually, as her attitude continued to deteriorate, the church-sponsored financial aid would diminish, then cease, and she finally left the church greatly offended. A week or two later, she showed up at another open-hearted church ... and the process of exploitation started all over again.

There are endless variations on this theme. These "beggar-believers" have entirely missed the principle of "sowing out of your own need"! They have become distracted by their problems so much that they've forgotten to keep their eyes on God – after all, ***He*** is their Source ... not welfare, not alms, not "lucky breaks."

Good Honest Labor

I looked it up: there are 52 direct Bible references to good honest work (most of them in the five books of Moses). A strong work ethic is a strong Christian attribute. Industry, forethought, careful stewardship, generosity – these are all honorable traits; read about the *"virtuous woman"* of Proverbs 31, and the *"five wise and five foolish virgins"* of Matthew 25. *"The laborer is worthy of his hire"* (Luke 10:7b).

The Widow of Zarephath was out gathering firewood because she was willing to work right up to her last meal, even her last breath in life. (It is my own invention that she was once a wealthy land-owner, the "lady of the house," that gathering firewood was probably previously not her responsibility, therefore "beneath her dignity." I have also known "wealthy" people to "get down and dirty" when it comes to being in a survival mode.)

What else did she have in her hand? I don't suppose many would consider a handful of flour and a few drops of oil worth much of anything. To the Widow, it meant a final mouthful for her son, a last meal for them both. As a mother myself, I expect I would gladly give my last bite to my child, foregoing that nourishment for myself – although I would probably lick my fingers!

To the Prophet of God – who had recently been miraculously sustained by raven-brought bread and meat – this same handful of flour and drops of oil represented **"potential abundance"** ... when given to God in sacrifice. When Elijah heard her recitation of pitiful assets (see 1 Kings 17:12), he knew he had found the key to provision for her and himself too.

He spoke, *"Make me a little bread cake from it **first,** and bring it out to me"* (1 Kings 17:13a). As God's representative, he was instructing her to give to God **first.**

How many times do you have to learn and re-learn this lesson: ***God comes first in our lives*** (Exodus 20:2-3). It is the basic and first law of living, but our ordinary lives – with the trials and problems we continually walk through – are **always** trying to distract our focus.

I Expect A Miracle

I anticipate the inevitable
Supernatural invention of God
I expect a miracle, I expect a miracle
I expect a miracle

I know God is my Source, He's the strength of my life
And when I'm planting seeds, He will honor His Word
I expect a miracle, I expect a miracle
I expect a miracle

It's the principle of sowing our seeds
That will activate all the blessings of God
I expect a miracle, I expect a miracle
I expect a miracle.

– Judy A. Gossett and Jeanne Halsey

So God, through His Prophet Elijah, tested the Widow by asking her to give up her "last seed" (the remaining flour and oil). Note I said **ask.** God had earlier said, *"I have **commanded** a widow-woman to sustain you"* (1 Kings 17:9), because He had set His Word into motion. But Elijah **asked** the Widow to bring him water and bread (see 1 Kings 17:10-11) because **we** can (temporarily) thwart the fulfillment of God's Word by our wills – by our receptiveness (or bad attitude), our action (or inaction), our open hearts (or closed minds) to the realm wherein God operates, by our agreement (or our aversion) to walking by faith.

Satan Takes Advantage

Initially the Widow protested, thinking the Prophet did not understand the severity of her circumstances. She responded with a recitation of evident facts: the long drought, the last of her food, the inevitable unwilling suicide. Perhaps she intended to feed her little son one last time, and then, when he was asleep, tragically to suffocate him, putting an end to his suffering from starvation. Then perhaps she would slit her own wrists or hang herself, or somehow bring her once-wonderful life to a sad close.

This kind of logic does not come from God. There is a verse that illustrates a timeless truth: *"Be of sober spirit; be on the alert. Your adversary, the devil, prowls about like a roaring lion, seeking someone to devour"* (1 Peter 5:8). While it is true the drought came as a result of the word spoken by the Prophet Elijah (on behalf of the Lord God, to the wicked king Ahab; see 1 Kings 17:1) as part of a call to repentance nation-wide, still Satan was taking full advantage of the vulnerability of the people – their sense of loss, disappointment, despair, even stirring up anger against God – *"to steal, kill and destroy"* (see John 10:10).

Anywhere the destructive work of Satan is evident, the stench of his presence lingers – the smell of fear. So immediately the Prophet countered her negativity with *"Fear not!"* (1 Kings 17:13). He was not insensitive to her plight but knew he had to quickly eradicate the dominance of Satan, to realign her vision with God's perspective.

I don't really think the Widow was suicidal, but I do think she had come to the last of her resources, the literal bottom of her barrel, the end of that hopeless rope. She genuinely saw no alternative to death. She was without hope, she was at the end of herself ... and that is where God can **begin** His most effective work in us:

> *Therefore I am well content with weaknesses, with insults, with distresses, with persecution, with difficulties, for Christ's sake; for when I am weak, then I*

> *am strong.*
>
> – 2 Corinthians 12:10

Just like Moses was outcast in the deserts of Midian, this Widow was at the place of supreme humility.

The Value of Humility

Humility is a really very under-valued concept in our world today. We tend to equate humility with debasement, diminishment, with loss and embarrassment:

> *But He gives a greater grace. Therefore it says, "God is opposed to the proud, but gives grace to the humble."*
>
> – James 4:6

> *And all of you, clothe yourselves with humility toward one another, for God is opposed to the proud but gives grace to the humble.*
>
> – 1 Peter 5:5b

Have you ever watched a nude toddler playing? That child is not ashamed of his nakedness, is not concerned if he will ever feed himself, doesn't think about amassing a fortune or leading a nation. He is a humble child, confident in his parents' card. That attitude of child-like faith – *"Whosoever then humbles himself as this child, he is the greatest in the Kingdom of Heaven"* (Matthew 18:4) – is **exactly** what God looks for in His children when He wants to do marvelous things.

"Humble" means a return to that simple, pure place of fellowship which Adam and Eve enjoyed with God so long ago in the Garden of Eden, before they went on their ill-fated "fruit-finding mission," when they were both *"naked and not ashamed"* (see Genesis 2:25).

Take the true definition of having a humble heart filled with child-like faith ... and strip off all our pedigrees, our self-importance, our pride, and come into the place of God's presence with *"clean (naked) hands and a pure (humble) heart"* (see Psalm 24:4). Even Paul, the greatest apostle, knew the value in stripping away his "rights" and declaring himself worthy only to *"boast in the Lord"* (see 2 Corinthians 11:21-30). It is better to cultivate a humble, child-like heart as a way of life than to have that ultimately "forced" on us – *"Obedience is better than sacrifice"* (1 Samuel 5:22) – as did the circumstances of the lives of Moses and the Widow of Zarephath.

I Want to Drink From the River of Life

(Chorus) *I want to drink from the River of Life*
I want to live an dwell in His presence
I want to drink from the River of Life
Fill me with fresh oil, fill me with wine

(Verse 1) *Yesterday's bread won't feed me today*
Yesterday's prayer's not what my heart wants to say
I want to know You better each day
Fill me with fresh oil, fill me, I pray

(Verse 2) *Break down my pride, my fear and my pain*
In the core of my heart, revive me again
Pour out Your power, shower Your rain
Fill me with fresh oil, fill me again.

– Ty Tyler and Jeanne Halsey

God tested the Widow, He moved her from a state of fear and lack into a place of hope and expectation. She responded ... and He more than supplied her needs. *"My*

God shall supply all your need, according to His riches in glory by Christ Jesus" (Philippians 4:19).

Chapter 5: **Bread and Fish**

"O*kay, Jeanne, that's all Old Testament stuff. Show me how it works in the New Testament."* I'm so glad you asked! One of the few events in the Bible consistently reported by all four Gospel writers is ***"the Feeding of the Five Thousand."*** This creative miracle caught the attention of Matthew (recorded in Matthew 14:15-21), Mark (in Mark 6:34-44), Luke (in Luke 9:12-17), and John (see John 6:1-14). Because I think it's important to study the nuances of these well-known parallel stories to better understand God's principle of provision, let's read them one after another.

Matthew's Story

And when it was evening, the disciples came to Him saying, "The place is desolate, and the time is already past; so send the multitudes away, that they may go into the villages and buy food for themselves."

But Jesus said to them, "They do not need to go away; you give them something to eat!"

And they said to Him, "We have here only five loaves and two fish."

And He said, "Bring them there to Me." And ordering the multitudes to recline on the grass, He took the five loaves and the two fish, and looking up toward Heaven, He blessed the food; and breaking the loaves He gave them to the disciples, and the disciples gave to the multitudes; and they all ate, and were satisfied.

And they picked up what was left over of the broken pieces, twelve full baskets. And there were about five thousand men who ate, aside from women and children.

– Matthew 14:15-21

Mark's Version

And when He went ashore, He saw the great multitude, and He felt compassion for them because they were like sheep without a shepherd; and He began to teach them many things. And when it was already quite late, His disciples came up to Him and began saying, "The place is desolate and it is already quite late; send them away so that they may go into the surrounding countryside and villages and buy themselves something to eat."

But He answered and said to them, "You give them something to eat!"

And they said to Him, "Shall we go and spend two hundred denarii on bread and give them something to eat?"

And He said to them, "How many loaves do you have? Go look!"

And when they found out, they said, "Five and two fish."

And He commanded them all to recline by groups on the green grass. And they reclined in companies of hundreds and of fifties. And He took the five loaves and the two fish, and looking up toward Heaven, He blessed the food and broke the loaves and He kept giving them to the disciples to set before them; and He divided up the two fish among them all.

And they all ate and were satisfied. And they picked up twelve full baskets of the broken pieces, and also of the fish. And there were five thousand men who ate the loaves.

– Mark 7:34-44

Luke's Report

And the day began to decline, and the twelve came and said to Him, "Send the multitude away, that they may go into the surrounding villages and countryside and find lodging and get something to eat; for here we are in a desolate place."

But He said to them, "You give them something to eat!"

And they said, "We have no more than five loaves and two fish, unless perhaps we go and buy food for all these people." (For there were about five thousand men.)

And He said to His disciples, "Have them recline to eat in groups of about fifty each." And they did so, and had them all recline. And He took the five loaves and the two fish, and looking up to Heaven, He blessed them, and broke them, and kept giving them to the disciples to set before the multitude.

And they all ate and were satisfied; and the broken pieces which they had left over were picked up, twelve baskets full.

– Luke 9:12-17

As John Saw It

After these things Jesus went away to the other side of the Sea of Galilee (or Tiberias). And a great multitude was following Him, because they were seeing the signs which He was performing on those who were sick. And Jesus went up on the mountain, and there He sat with His disciples.

Now the Passover, the feast of the Jews, was at hand. Jesus therefore lifting up His eyes, and seeing that a great multitude was coming to Him, said to Philip, "Where are we to buy bread, that these may eat?" And this He was saying to test them; for He Himself knew what He was intending to do.

Philip answered Him, "Two hundred denarii worth of bread is not sufficient for them, for everyone to receive a little."

One of His disciples, Andrew, Simon Peter's brother, said to Him, "There is a lad here who has five barley loaves and two fish, but what are these for so many people?"

Jesus said, "Have the people sit down." Now there was much grass in the place. So the men sat down, in number about five thousand. Jesus therefore took the loaves; and having given thanks, He distributed to those who were seated; likewise also of the fish as much as they wanted.

And when they were filled, He said to His disciples, "Gather up the leftover fragments that nothing may be lost." And so they gathered them up, and filled twelve baskets with fragments from the five barley loaves, which were left over by those who had eaten.

When therefore the people saw the sign which He had performed, they said, "This is of a truth the Prophet Who is to come into the world."

– John 6:1-14

How Marvelous!

Jesus Christ is so marvelous in this story! His tenderness and compassion for the people is so like Him! For three days – despite His personal anguish over the recent news that His beloved cousin, John the Baptist, had just been beheaded by King Herod – He has been pouring out of Himself freely, feeding the common people spiritually, healing them physically, strengthening them emotionally, challenging them intellectually with radical new ideas about their relationship to God and their relationships with their fellow human beings. And He was preparing to do the ultimate for them and all of Mankind: He was preparing to offer Himself as the final Passover Lamb (see John 6:4).

He cared about their well-being. There's an interesting little "aside" the writer John includes: *"And this He was saying to test them,* ***for He Himself knew what He was intending to do"*** (John 6:6). When believers are smack against a brick wall, their lives seemingly shattered, it is so important to remember: **nothing** ever catches God by surprise.

He Knows My Name

(Verse 1) *I have a Father*
He calls me His own
He'll never leave me
No matter where I go

(Chorus) *He knows my name*
He knows my every thought
He sees each tear that falls
And hears me when I call

And hears me when I call

(Verse 2) *I have a Maker*
He formed my heart
Before even Time began
My life was in His hands

(Chorus) *He knows my name*
He knows my every thought
He sees each tear that falls
And hears me when I call
And hears me when I call.

– Tommy Walker

Jesus Asked "The Question"

The first thing that jumped out to me about these four parallel accounts of the same event is that, in each version, Jesus essentially asked His disciples, ***"What's that you have in your hands?"***

But Jesus said to them, "They do not need to go away; ***you*** *give them something to eat!"*

– Matthew 14:16

*But He answered and said to them, "**You** give them something to eat!"*

– Mark 6:37a

*But He said to them, "**You** give them something to eat!"*

– Luke 9:13a

Jesus therefore lifting up His eyes and seeing that a great multitude was coming to Him, said to Philip, "Where are we to buy bread, that these may eat?" And this He was saying to test them; for He Himself knew what He was intending to do.

– John 6:5-6

By the way, because I am often accused of being overly dramatic, I want to point out that the exclamation points are added by the publishers of *The New American Standard Bible,* not by me.

The Principle of Multiplication

We all know this four-times-told miracle as being an illustration of the principle of supernatural multiplication. We know Jesus produced the thousands and thousands of pieces of bread, and thousands and thousands of slices of fish needed to feed *"about five thousand men who ate, aside from women and children"* (Matthew 14:21). But before He ever did anything, He wanted to test what the disciples understood about multiplication, to discover if they had learned anything about using "that which was already within their hands" to miraculously change the situation. He specifically asked them, *"What do you have on hand that can satisfy this dilemma and bring glory to God?"*

In reading the four parallel accounts, I see this development: the disciples were glad Jesus was so successful in His ministry, but they were looking at the huge crowd of people and starting to get worried. They could see this "idyllic setting of thousands of people rapturously hearing the wonderful teachings of the Master" which could quickly turn ugly when the crowd started to be hungry for natural food.

I have to give the disciples credit for "looking ahead" for once, instead of always just "living in the moment." They were only using their "flesh eyes" instead of "spirit eyes" to see the reality. "Flesh eyes" are usually accompanied by attitudes of worry, fear, doubt, anxiety, despair, hopelessness. "Spirit eyes" are usually marked by attitudes of anticipation, expectation, *"unspeakable joy,"* and a readiness to give overflowing glory to God.

Isn't it too easy for Christians to get side-tracked by details and miss the beauty of God's handiwork? *"Our church just entered into a $2.1-million building program" ...* or *"We've planted sister churches in Hispanic and Korean neighborhoods, and started an off-campus fellowship next to the university"* ... or *"We just had our largest attendance ever, and have now moved to three Sunday morning services."* These are relatively "good" things – but never let these "things" steal the glory from God, Who alone deserves it: *"When therefore the people say the sign which He had performed, they said, 'This is of a truth the Prophet Who is to come into the world'"* (John 6:14). If "things" bring glory to Man, then they are misguided. I would rather be part of a church that says, *"Holy Spirit showed up in the most remarkable way this week!"*

The Disciples' *Modus Operandi*

The basic response of the disciples is disappointingly familiar: **doubt.** It isn't very encouraging for the average Christian today -- who doesn't have their incomparable experience of actually walking and talking with Jesus Christ in the flesh, living with Him, learning from Him, personally witnessing His wondrous miracles – to realize that they still had doubts! Either they were, as a group, incredibly dense or ... ***wait a minute!***

They did not yet have what **we** have today: the indwelling Holy Spirit of Jesus Christ, with us every moment, every minute of every hour, as close as our next breath, as near as our next heartbeat! So we'll just cut these guys some slack because, maybe, they were just *too* close to the Tree of Life to see the whole forest.

Jesus Held the Key

Have you ever lost your keys? If you live with people who like to play "practical jokes" (like some in my family do, bless their hearts), have your lost keys ever been found by someone else who "kept" them for awhile just to tease or aggravate you? I'm not implying that Jesus was teasing His disciples when He asked them a "rhetorical" question. I believe Jesus was trying to get them to actually **see** what they had on hand to work with ... and because He was signaling to His disciples that He was very aware of the situation, and He was preparing to do something about it.

Look at the way Mark wrote the story:

But He answered and said to them, "You give them something to eat!"

And they said to Him, "Shall we go and spend two hundred denarii on bread and give them something to eat?"

And He said to them, 'How many loaves do you have? Go look!"

– Mark 6:37-38

Jesus was not vague about what He was going to feed the multitude – He was quite specific: because He knew the little boy was there ... because He knew the food was there ... because He knew what kind of bread it was ... because He knew what kind of fish it was. But the disciples were having a hard time seeing these things. Their "flesh eyes" were so limited – were they looking for some sort of *"Pita-To-Go"* or some sort of catering truck (or chariot) to drive up and open up for business? And only John was clued-in enough to acknowledge the little boy's participation in the whole event!

Jesus Himself held the key to the Bank of Heaven, and He knew how to release the unlimited resources from His Father's Storehouse! "To buy" means to give something in exchange for another thing; a transaction. As in putting a coin in the machine and getting a bottle of soda pop in return. Here Jesus was again presenting the oft-repeated lesson of sowing and reaping, of giving to God your insignificant seed to unlock the flood of Heaven's blessing.

What Kind of Seed?

The dilemma (the way the disciples saw it): *"We just don't have enough to feed these people."* In John's report, it was Philip who answered, *"Two hundred denarii is not enough ... for five thousand men"* (John 6:7, 10; paraphrased). Can't you just see Philip pulling out the Disciples' Treasury Check-book and looking at their account balance, adding and re-adding the figures in case someone had made a mistake ... and still coming up with a measly two hundred denarii. (One denarius was generally considered one day's wage, so two hundred denarii was substantial.)

How often we see what we don't have! Why are we so inclined to see the glass of water "half-empty" rather than "half-full"? Whatever the balance in the Disciples' Treasury may have been, Jesus was not interested in that, anyway. He wasn't interested in their money – He was interested in their **seed.**

Jesus asked, "How many loaves do you have? Go look!"

– Mark 6:38

Andrew answered, "There is a lad here who has five barley loaves and two fish, but what are these for so many people?"

– John 6:9

The Attitude of Expectation

In all four versions, Jesus then said, *"Have the people sit down"* (see Matthew 14:19; Mark 6:39; Luke 9:14-15; John 6:10). This was to put the people into a calm, relaxed attitude of expectation. It was orderly and thorough, there was nothing hysterical about it. People could clearly see Him taking a small amount of food ... and filling basket after basket of supply for everyone! There was no sleight-of-hand taking place, no one sneaking food in through the confusing of a milling crowd. They accepted His food with gladness ... much the same way the Israelites had gathered manna in the wilderness (see Exodus 16).

In one of the better movies of 1977, brilliant director Franco Zeffirelli wrote and filmed *"Jesus of Nazareth."* One of my favorite scenes in the entire six-hour production is this very story of "Feeding the Five Thousand." A sweet twist has "Mary Magdalene" (wonderfully acted by Anne Bancroft), a cynical prostitute, coming to hear the radical teachings of this "strange Rabbi from Galilee" (dynamically portrayed by Robert Powell). She is anonymous among the crowd, and hears His teachings with a certain worldly reserve. But when she is handed a beautiful loaf of fresh bread, she sort of absently bites into it ... and suddenly realizes she has witnessed and partaken in a miracle!

Tears overflow her expressive eyes ... and the bitter heart of this fallen woman is changed forever. Such a poignant scene.

The Little Boy's Story

But, let us not forget the little boy whose home-packed lunch was the humble seed Jesus accepted to initiate Heaven's harvest. For so many of us, it is too easy to overlook the actual assets we hold in search of some flashier sign. Here's the way I related this Biblical to my young grandson Kristian, one night at a bed-time story:

Joey was a little boy who lived in Bible times. One day, he wanted to go on a picnic, so his mother packed his lunch-bag with his favorite food – barley bread sticks and sweet-pickled fish – and sent him on his way. While Joey was playing and wandering through the meadows and the woods, he came across a very unusual scene: a huge crowd of people had gathered all around a hill to hear a famous traveling Teacher speak.

Joey was just a little boy and could easily slip through the crowd, so he worked his way right to the top of the hill and sat down very closely to listen to this Man's teaching. He liked the Teacher so much that he stayed all day, just to listen. Pretty soon, it got to be lunch-time, and Joey decided to open his picnic lunch.

Then Joey overheard the Teacher's helpers talking. They were worried about needing food for all the people. Joey looked at his little lunch-bag, then at the Teacher's helpers, then at his lunch-bag again. Finally, he stood up and wiggled over to one of the Teacher's helpers. "Mister, excuse me," he said politely, pulling on the helper's sleeve.

The helper looked down at the little boy. "Huh ... what is it, kid?"

"Do you say you need food?"

"Just a minute, kid. We're kind of busy here!" The helper started to turn away.

"But I have food!" Joey insisted.

The helper just stared at Joey. "What do you mean, kid?"

Joey held up his little lunch-bag. "I have food!" he repeated. The Teacher's helper slowly smiled. "And I would like to give mine to the Teacher," Joey added. So Joey handed over his little lunch-bag to the Teacher's helper, who then gave it to the Teacher.

Joey watched the Teacher reach into his little lunch-bag and pull out one of his barley bread sticks. He held it up and said, "Father, thank You for this food." Joey was sort of surprised, and wondered what would happen next.

Then the Teacher snapped the bread stick in half ... and with His right hand, handed a whole bread stick to His helper while holding the half-stick in His left hand. Then the Teacher snapped the whole bread stick in His left hand in half again, and again He handed ***another*** *whole bread stick to another helper. Joey watched as the Teacher kept breaking the same whole bread stick over and over again, and food was being passed along to all the people who were there – and there were five thousand men, plus women and children! The Teacher* *just never ran out of bread.*

Then Joey saw the Teacher reach into his little lunch-bag and pull out a sweet-pickled fish, which He handed to a helper who passed it along to others. Again and again, the Teacher pulled sweet-pickled fish out of his little lunch-bag, until there was more than enough to feed all the people. After awhile, the helpers gathered up all the leftover food, and there were twelve huge baskets filled with Joey's lunch!

Joey was so proud. His little lunch-bag had become super-special because the Teacher was feeding everyone from it. Then the Teacher called for Joey to come close. "Thank you, little boy, for sharing your lunch with all of us." He handed the little lunch-bag back to Joey ... and when he looked inside, he found a chocolate bar (which had not yet been invented)! The End.

(Kristian really liked that silly part at the end!)

Chapter 6: THE IMPORTANCE OF THANKSGIVING

As I studied the topic of thanksgiving – which is covered extensively in the Word, especially among the Psalms of David – I was reminded of a conversation I once had with a distinguished linguist He had studied professionally all the languages of the world, including ancient "dead" languages that have given rise to modern ones. He said in all the languages he had studied, there were basic common nouns for *man ... woman ... child ... fire ... water ... dark ... light ...* and more. Then he said there were common verbs for *stand ... walk ... run ... sit ... sleep ... eat ... give ... take ...* and so on.

He continued, "All languages have words that deal with more abstract concepts, such as *build ... hunt ... trade ... share* – but not all languages have a word for the somewhat obscure concept of *thanksgiving.* There are some ancient cultures which have completely died out, have not spawned any direct descendants ... but in their languages, I cannot find the words corresponding with *thanks* or *thanksgiving.* They did not acknowledge a 'higher authority' in their culture, they were not courteous with each other or outsiders. They were fierce, war-like, 'every man for himself' cultures, and they eventually killed themselves into extinction."

> *But realize this, that in the last days difficult times will come. For men will be lovers of self, lovers of money, boastful, arrogant, revilers, disobedient to parents, ungrateful, unholy, unloving, irreconcilable, malicious gossips, without self-control, brutal, haters of good, treacherous, reckless, conceited, lovers of pleasure rather than lovers of God.*
>
> – 2 Timothy 3:1-4

Jesus Christ, the Son of God, the Descendant of King David, was thankful. He always pointed to His Heavenly Father as His Source, and He taught people to be thankful – especially **before** the fact. Jesus equated "giving thanks to god" with "giving *glory* to God" (see Luke 17:11-19) – *"Was no one found who turned back to give glory to God, except this foreigner?"* (Luke 17:18).

Genuine Thanksgiving *Precedes* Receiving

Perhaps the most dramatic account of Jesus giving thanks to God before the manifestation of the miracle is found in the story of the resurrection of Lazarus in John 11:

> *Now a certain man was sick, Lazarus of Bethany, the village of Mary and her sister Martha. And it was the Mary who anointed the Lord with ointment, and wiped His feet with her hair, whose brother Lazarus was sick. The sisters therefore sent to Him, saying, "Lord, behold he whom You love is sick."*
>
> *But when Jesus heard it, He said, "This sickness is not unto death, but for the glory of God, that the Son of God may be glorified by it."*
>
> *Now Jesus loved Martha, and her sister, and Lazarus. When therefore He heard that he was sick, He stayed then two days longer in the place where He was. Then after this He said to the disciples, "Let us to go Judea again. ... Our friend Lazarus has fallen asleep; but I go, that I may awaken him out of sleep."*
>
> *The disciples therefore said to Him, "Lord, if he has fallen asleep, he will recover." Now Jesus had spoken of (Lazarus') death, but they thought that He was*

speaking of literal sleep.

Then Jesus therefore said to them plainly, "Lazarus is dead, and I am glad for your sakes that I was not there, so that you may believe; but let us go to him." ...

So when Jesus came, He found that he had already been in the tomb four days. Now Bethany was near Jerusalem, about two miles off; and many of the Jews had come to Martha and Mary, to console them concerning their brother. Martha therefore, when she heard that Jesus was coming, went to meet Him; but Mary still sat in the house Martha therefore said to Jesus, "Lord, if You had been here, my brother would not have died. Even now I know that whatever You ask of God, God will give you."

Jesus said to her, "Your brother shall rise again."

Martha said to Him, "I know that he will rise again in the resurrection on the last day."

Jesus said to her, "I am the Resurrection and the Life; he who believes in Me shall live even if he dies, and everyone who lives and believes in Me shall never die. Do you believe this?"

She said to Him, "Yes, Lord, I have believed that You are the Christ, the Son of God, even He Who comes into the world." ...

Therefore, when Mary came where Jesus was, she saw Him and fell at His feet, saying to Him, "Lord, if You had been here, my brother would not have died."

When Jesus therefore saw her weeping, and the Jews

who came with her also weeping, He was deeply moved in spirit and was troubled, and said to them, "Where have you laid him?"

They said to Him, "Lord, come and see." Jesus wept. ...

Jesus therefore again being deeply moved within, came to the tomb. Now it was a cave, and a stone way lying against it. Jesus said, "Remove the stone."

Martha, the sister of the deceased, said to Him, "Lord, by this time there will be a stench, for he has been dead four days."

Jesus said to her, "Did I not say to you, if you believe, you will see the glory of God?" And so they removed the stone. And Jesus raised His eyes and said, "Father, I thank Thee that Thou heardest Me. And I knew Thou hearest Me always; but because of the people standing around I said it, that they may believe that Thou didst send Me."

And when He had said these things, He cried out with a loud voice, "Lazarus, come forth." He who had died came forth, bound hand and foot with wrappings, and his face was wrapped around with a cloth. Jesus said to them, "Unbind him, and let him go."

Many therefore of the Jews, who had come to Mary and beheld what He had done, believed in Him.
– John 11:1-7, 11-15, 17-27, 32-35, 38-45; paraphrased

Note that Jesus knew **exactly** what was going on at all times. Let us suppose it was a Monday when Jesus got the message that Lazarus was sick (and dying). The first thing He said was, *"This sickness is not unto death, but for the* *glory of God, that the Son of God may be glorified by it"* (John 11:4). He wasn't callous about Lazarus and his family; He loved them dearly – *"Now Jesus loved Martha, and her sister, and Lazarus"* (11:5) – but He loved His Father even more, and loved bringing glory to Him even more. So He did not get up and immediately leave that place to rush over to Bethany ... He stayed until Wednesday.

He knew, even as the message reached Him on Monday, that Lazarus had died – by the time He eventually reached Bethany on Thursday, Lazarus had been dead four days (see 11:17, 39). He was not delaying because He was "sorting through His bad of tricks" nor because He was puzzling over how to "solve this mess" (see 11:39), as some of His detractors (past and present) claimed. Jesus lingered, not to increase the loss and grief that His beloved friends Mary and Martha were experiencing, but to increase the glory which would be given to God (see 11:4, 20).

It's not like when He heard the news on Monday, there was still hope He could heal Lazarus before he succumbed to death. After all, the message from Mary and Martha was probably sent on Saturday or Sunday, taking at least a full day in transit ... and by the time it reached Jesus on Monday, Lazarus had already died, and a miracle-healing was already too late. So it was not a matter of logistics gone wrong – Jesus simply knew that in raising Lazarus from the dead, God would receive greater glory than could be generated by a simple "rise up and be healed" miracle (see 11:15).

The crux comes when Jesus stands before the tomb of his dear friend and says to Martha, *"Did I not say to you, if you believe, you will see the glory of God?"* (11:40). Then the men opened the grave at His command, and ***He gave***

thanks to God first: *"And Jesus raised His eyes and said, 'Father, I thank Thee that Thou heardest Me. And I knew that Thou hearest Me always; but because of the people standing around I said it, that they may believe that Thou didst send Me'"* (11:4-42).

Note that Jesus could have done all His praying "inside His heart" – that is, His relationship with His Father was so intense that verbalizing aloud was not actually necessary. However, He prayed aloud so that people all around could hear – and learn from – His method of praying, **including starting the prayer with thanksgiving.**

Thanksgiving is *Not* a Bribe

His thanksgiving was not hocus-pocus or some magic incantation. I think it is good to teach our small children the concept of thankful prayers – *"Now I lay me down to sleep / I pray the Lord my soul to keep / When I awake to morning light / I thank the Lord for restful night"* and *"God is great, God is good / Now we thank Him for our food"* and *"Bless this food to the nourishment of our bodies"* – but soon they should be encouraged to pray and give thanks spontaneously and genuinely from their own hearts, in their own words.

"And Jesus raised His eyes and said, 'Father, I thank Thee that Thou heardest Me" (11:41). *"Heardest"* is past-tense – Jesus knew God **already** knew about the situation, had **already** reversed the decaying process of death in Lazarus' body. Jesus and the Father have **always** been "Partners in Miracles": *"Hence also He is able to save forever those who draw near to God through Him, since He always lives to make intercession for them"* (Hebrews 7:25).

Jesus glorified God: *"And I knew that Thou hearest*

Me always; but because of the people standing around I said it, that they may believe that thou didst send Me" (11:42). Jesus glorified God when He fed the multitudes. Jesus was not "lucky" or "brought good luck to others" – He was in active partnership with God to bless people and ***to bring glory to God.*** Jesus exemplified the "attitude of gratitude."

Chapter 7: **SILVER AND GOLD**

Thus far we have discussed the principle of tangibly using what you have in your own hands to fulfill the purposes of God in your life. But now consider the "intangible":

> *Now Peter and John were going up to the Temple at the ninth hour, the hour of prayer. And a certain man who had been lame from his mother's womb was being carried along, whom they used to set down every day at the Gate of the Temple which is called Beautiful, in order to beg alms of those who were entering the Temple.*
>
> *And when he saw Peter and John about to go into the Temple, he began asking to receive alms. And Peter, along with John, fixed his gaze up him and said, "Look at us!"*
>
> *And he began to give them his attention, expecting to receive something from them. But Peter said, "I do not possess silver and gold, but what I do have I give to you: in the Name of Jesus Christ the Nazarene – walk!"*
>
> *And seizing him by the right hand, he raised him up; and immediately his feet and his ankles were strengthened. And with a leap, he stood upright and began to walk; and he entered the Temple with them, walking and leaping and praising God.*
>
> *And all the people saw him walking and praising God; and they were taking note of him as being the one who used to sit at the Beautiful Gate of the Temple to beg alms, and they were filled with wonder and amazement at what had happened to him.*

– Acts 3:1-10

Almost every child in Sunday School learns this little song: *"Silver and Gold have I none / But such as I have, give I thee / In the Name of Jesus Christ / Of Nazareth, rise up and walk!"*

A "Police Report"

Imagine you are a police detective doing background research for a "disturbing the peace" investigation. Here's what you might write up in your case report for your superior officer:

Case #: *3-1-10*
Subject (full name): ***SIMON PETER BAR JONAH*** *a/k/a "the Rock"*
Home Address: *Hotel Arimathea, downtown Jerusalem (formerly Capernaum in Galilee)* **Business Address:** *Upper Room, Hotel Arimathea*
Status: *Married; one son*
Age: *Approximately 40*
Description: *Tall, heavy-set, brawny; fisherman by trade*
Estimated Taxable Financial Worth: *Unknown; presumed none*
Background History: *Subject once owned a lucrative family fishing business with his brother Andrews and partner Zebedee (father of James and John), all of Galilee. Over 3 years ago, subject unexpectedly abandoned his business to become a follower of an itinerant preacher, known as "Jesus of Nazareth" (now deceased; executed for heresy) (see associated file: "Corpse stolen from sealed tomb").*
Subject appears to have no visible means of employment, depending on the charity and goodwill of approximately 1,000 to 3,000 citizens, who are

forsaking acceptable standard religious practice in "celebration" of the alleged resurrection of Jesus of Nazareth (note prior reference). After the death of their cult's founder, subject has become self-appointed leader and primary teacher of this new religion; he is not a great orator and has been known to be publicly offensive, but he does have remarkable results in gathering new "converts."

Estimated Danger Quotient: ***Extremely high.*** *Keep under constant surveillance*

Case Status: *Active*

Case #: *3-1-11*

Subject (full name): ***JOHN BEN ZEBEDEE***

Cross-Reference: *SIMON PETER BAR JONAH*

Home Address: *Hotel Arimathea, downtown Jerusalem (formerly Capernaum in Galilee)* **Business Address:** *Unknown*

Status: *Unmarried*

Age: *Approximately 30*

Description: *Average height, sturdy build; fisherman by trade*

Estimated Taxable Financial Worth: *None*

Background History: *Previously employed as a fisherman with "Bar Jonah, Zebedee & Sons" in Galilee, subject also abandoned career and business interests to follow controversial itinerant preacher Jesus of Nazareth. A constant companion of Simon Peter, John is less public although he is "revered" as a so-called "apostle" among the new religionists. It is rumored that subject is writing an account of the life and ministry of Jesus of Nazareth.*

Estimated Danger Quotient: ***Fairly high.*** *Associated surveillance-mode*

Case Status: *Active*

Case #: *4-13-07*
Subject (full name): ***REUBEN, the Beggar***
Home Address: *Unknown.* **Business Address:** *The Gate Beautiful, Jerusalem Temple*
Status: *Unmarried*
Age: *Approximately 50*
Description: *Lame from birth. Beggar by trade*
Estimated Taxable Financial Worth: *None*
Background History: *Licensed by Temple authorities to beg at the Gate Beautiful.*
Estimated Danger Quotient: ***Harmless, but noisy***
Case Status: *Pending*

Peter and John Were "Ordinary Guys"

Okay, enough *"Law & Order."* The point is: Peter and John were poor by the world's standards. It was well over three years since they had *"left all to follow Jesus"* (see Matthew 4:20-21); who knows what they lived on during those years? But after all the wonders and life-transformations on the Day of Pentecost, they were so consumed with God's grace and glory that they were rich beyond measure!

> *And they were continually devoting themselves to the apostles' teaching and to fellowship, to the breaking of bread and to prayer. And everyone kept feeling a sense of awe; and many wonders and signs were taking place through the apostles.*
>
> *And all those who had believed were together, and had all things in common; and they began selling their property and possessions, and were sharing them with all, as anyone might have need. And day by day continuing with one mind in the Temple, and breaking bread from house to house, they were taking their*

meals together with gladness and sincerity of heart, praising God, and having favor with all the people.

And the Lord was adding to their number day by day those who were being saved.

– Acts 2:42-47

It is true the San Hedrin was probably still upset with them, and the Romans were probably hoping they would just fade away. But Peter and John were enthused – meaning "filled with God" – and their enthusiasm was contagious. They were having daily revival services in the Temple, and despite their shortcomings in education and articulation, their unpolished speech – *"Now as they observed the confidence of Peter and John, and understood that they were uneducated and untrained men, they were marveling and began to recognize them as having been with Jesus"* (Acts 4:13) – they were seeing daily conversions among Jews and Gentiles both.

"Reuben the Beggar"

"Reuben the Beggar" is the made-up named I've given this particular man mentioned in Acts chapter 3. He was even poorer in material goods and financial prospects – and certain poorer in physical health – than Peter and John. He was *"lame from birth,"* and in those days, he was pitied, given minimum care (had had family or friends who daily carried him to and from his crude lodgings; see Acts 3:2), but his society did not put much value on his life, and expected him to wither and die sooner rather than later.

Reuben was also something of a well-known "institution," even a popular attraction at the Temple. **Everybody** knew him – maybe he was a prototype "sit-down comic" (sorry about that, I didn't really mean to be corny) ... or perhaps a political satirist (see 3:10). In any case, every beggar I've ever met – and I've encountered them in every culture I've known around the world, even in these "enlightened" times – has always had some sort of gimmick or verbal "patter" to attract attention. By all indications in Acts 3, Reuben was probably somewhat personable, outspoken, even likable ... just permanently lame and a beggar, a man with no future beyond each day, no real hope of recovery, no real prospects or purpose in life.

The "Intangible" Difference

So when two not-wealthy men (Peter and John) encountered an even poorer man (Reuben), it was not promising in the vast exchange of wealth. ***But!***

> ***But*** *Peter said, "I do not posses silver and gold,* ***but*** *what I do have I give to you: in the Name of Jesus Christ of Nazareth – walk!*
>
> – Acts 3:6

Peter did not have anything tangible in his hands – no silver to buy a meal for Reuben, no gold to purchase a new cloak to cover his rags. Peter wasn't even an orthopedic surgeon or a massage therapist who could have used his hands to try to effect a cure or at least some sort of relief for Reuben's lame condition. So what did Peter have to give?

- **The authority of the Name of Jesus Christ**
- **The right as a believer to use that Name**
- **The power of Holy Spirit to heal the sick**
- **The willingness to be the vessel to conduct that power of Heaven right into the withered body of the beggar**
- **The boldness to contradict Satan's destructive**

handiwork in Reuben's life

- **And the courage to do so right in public, right on the steps of the Temple!**

That quality of boldness seems to be a characteristic important to the activation of God's supernatural blessing. Boldness was often attributed to Peter (see Acts 1:15, 2:14, 3:6-7, 12, 4:8-10, 13, 29-31). Boldness was also often attributed to Paul (formerly Saul of Tarsus) (see 2 Corinthians 7:4; Ephesians 3:12; Philippians 1:20; 1 Timothy 3:13; Hebrews 10:19), and to the other apostles and members of the Early Church.

"Boldness" is often defined as "great confidence." In any situation in life, the recognition that God is our Source, and the willingness to bring glory to Him, infuses us with that great confidence which enables us to release whatever it is we have in our hands into His keeping ... into His creative abundance, where our salvation lies and from where His blessings pour.

Peter did not have anything tangible in his hands, but he unleashed the healing power of Jesus Christ ... and Reuben **leaped up – immediately** strengthened, immediately healed, immediately coordinated enough to walk and leap (see 3:4-9), immediately grateful to God, and **loud** in his praises!

And because Peter was filled with boldness, he took the opportunity to speak to the amazed crowd standing around, preaching yet another soul-stirring salvation message (see 3:12-26) ... although this one landed him and John (and possibly Reuben too) in jail overnight (see 4:1-3).

That which you have in your hands can be released with gratitude into God's hands, with the

purpose of bringing great glory to Him, knowing fulfillment of His plan for your life, and receiving abundant blessings for you and your household, even for your nation.

Chapter 8: **The Ministry of Hands**

The Word is filled with references to our hands – over 107 references. Our hands are used to bless (see Deuteronomy 16:15, 24:19, 33:11; Psalm 63:4, 134:2; 1 Corinthians 4:12). Our hands are used to hurt, to wound, to kill (see Exodus 29:10-11, 19-20; Numbers 24:10-11; 1 Samuel 30:15; Matthew 17:22; Mark 9:31). Our hands can create ... our hands can destroy.

I especially enjoy the verses that speak of the **condition** of our hands: *"Who may ascend into the hill of the Lord? And who may stand in His holy place? He who has clean hands and a pure heart"* (Psalm 24:3-4) – and the **place** of our hands: *"Nevertheless I am continually with Thee; Thou hast taken hold of my right hand. With Thy counsel Thou wilt guide me, and afterward receive me into glory. Whom have I in Heaven but Thee? And besides Thee, I desire nothing on Earth. My flesh and my heart may fail, but God is the strength of my heart and my portion forever"* (Psalm 73:23-26) – and the **actions** of our hands: *"Give her the product of her hands, and let her works praise her in the gates"* (Proverbs 31:31). Ecclesiastes 9:10 is an inspiring verse: *"Whatever your hand finds to do, verily, do it with all your might."* Paul exhorts us to *"pray, lifting up holy hands"* (1 Timothy 2:8).

It is beautiful read about and visualize what Jesus Christ Himself did with His hands:

And behold, a leper came to Him, and bowed down to Him, saying, "Lord, if You are willing, You can make me clean."

And He stretched out His hand and touched him,

saying, "I am willing; be cleansed." And immediately his leprosy was cleansed. ...

And when Jesus had come to Peter's home, He saw his mother-in-law sick in bed with a fever. And He touched her hand, and the fever left her; and she arose. ...

And when evening had come, they brought to Him many who were demon-possessed; and He cast out the spirits with a word, and healed all who were ill, in order that what was spoken through Isaiah the prophet might be fulfilled, saying, "He Himself took our infirmities, and carried away our diseases."

– Matthew 8:2-3, 14-15, 16-17

While He was saying these things to them, behold, there came a synagogue official, and bowed down before Him, saying, "My daughter has just died; but come and lay Your hand on her, and she will live." And Jesus arose and began to follow him, and so did His disciples.

And behold, a woman who had been suffering from a hemorrhage for twelve years, came up behind Him and touched the fringe of His cloak; for she was saying to herself, "If I only touch His garment, I shall get well."

But Jesus turning and seeing her, said, "Daughter, take courage; your faith has made you well." And at once the woman was made well.

And when Jesus came into the official's house, and saw the flute-players, and the crowd in noisy disorder, He began to say, "Depart, for the girl has not died, but is asleep." And they began laughing at Him. But when the crowd had been put out, He entered and took her by the hand; and the girl arose. And this news went out into all that land.

– Matthew 9:18-26

Then some children were brought to Him so that He might lay His hands on them and pray; and the disciples rebuked them. But Jesus said, "Let the children alone, and do not hinder them from coming to Me; for the Kingdom of Heaven belongs to such as these." And after laying His hands on them, He departed from there.

– Matthew 19:13-15

While He was still speaking, behold, a multitude came, and the one called Judas, one of the twelve, was preceding them; and he approached Jesus to kiss Him. But Jesus said to him, "Judas, are you betraying the Son of Man with a kiss?"

And when those who were around Him saw what was going to happen, they said, "Lord, shall we strike with the sword?" And a certain one of them struck the slave of the high priest and cut off his right ear.

But Jesus answered and said, "Stop! No more of this." And He touched his ear and healed him.

– Luke 22:47-51

(Also see Mark 1:4, 31, 5:23, 9:27, 10:13, 16:18; Luke 18:15; John 13:3, 8:6, 9:11, 20:19; Acts 11:20-21)

Making Treasures

Many years ago, I contributed to a book written by my father, Rev. Don Gossett, called, *"If Nobody Reaches, Nobody Gets Touched."* One of the thing presented in the book was "the ministry of writing letters." In this technology-oriented "information age," few people take time to actually put pen to paper and hand write a letter to a friend.

I treasure certain letters or handwritten messages:

- Many, many letters of encouragement written in her beautiful script by my mother, Joyce Aletha Shackelford Gossett, who went to her Heavenly Reward in August 1991
- Several sweet childish scrawls by my children, Jennifer Elisabeth Joy and Alexander John Edward, when they were very young, with their letters so unevenly sized and trailing up and down the page
- A wonderful blessing upon his first great-grandson typewritten by my grandfather, Rev. William Canada Shackelford, upon the occasion of the birth of my son Alexander
- An exquisite floral printed card from my best friend, Judy Vanderhoof Godwin, as she shared her blend of emotions when she first became a mother
- A drawing of "something" by my grandson Kristian Michael Alexander Halsey, with his delightful, bold signature across the top
- A poignant birthday greeting from my sister Judith Anne Gossett, which truly described our indelible bond
- The print-out of a facsimile sent by my 19-year-old son Alex, as he was ministering in Thailand
- The loving note Jen left pinned to my pillow on the milestone day when she moved out of our home into her own
- The print-out of an e-mail received from my brother, Donnie Gossett, after we had been through a difficult, tumultuous experience together
- And so much more.

My scrapbooks and keepsake files are filled with letters I'd rather not lose or throw away. I just hope some of my letters, cards, notes, messages which I've sent to others throughout the years will have a fraction of the ongoing blessings which these mostly written-by-hand treasures have meant to me.

Let me encourage you to take on the "ministry of writing letters," not to become sentimental but to become a blessing. Unashamedly pour your heart into your writing ... for where would the world be today if we did not have the wealth of the heart of David the Psalmist, or Paul the learned Apostle, to read these words thousands of years later? And don't forget the "love-letter" written by the hand of Almighty God Himself on tablets of stone:

> *And when He had finished speaking with him upon Mount Sinai, He gave Moses the two tablets of the testimony, tablets of stone written by the finger of God.*
>
> – Exodus 31:18

> *And the Lord gave me the two tablets of stone written by the finger of God; and on them were the words which the Lord had spoken with you at the mountain form the midst of the fire on the day of assembly.*
>
> – Deuteronomy 9:10

Take advantage of modern technology to transmit your love and concern, your counsel and encouragement to others. *"The pen is mightier than the sword,"* wrote Cardinal Richelieu.

Take Hold

What is the ministry of hands? Technically, it means "to take hold of something." There are several word-pictures given in the Book of Genesis alone!

1. Hands *take* things:

> *Then the Lord God said, "Behold, the man has become like one of Us, knowing good and evil; and now, lest he stretch out his hand and take also from the Tree of Life and eat, and live forever" – therefore the Lord God sent him out east of the Garden of Eden.*
>
> – Genesis 3:22-23

When Adam and Eve sinned in the Garden of Eden, God mentioned the actions of Man's hands as having altered the order of things because he *"took hold with his hands"* that which he was not supposed to have.

2. Hands *make* things:

> *And his brother's name was Jubal; he was the father of who all who play the lyre and pipe. As for Zillah, she also gave birth to Tubal-cain, the forger of all implements of bronze and iron."*
>
> – Genesis 4:21-22

> *"Make for yourself an ark of gopher wood; you shall make the ark with rooms, and shall cover it inside and out with pitch." ... Thus Noah did, according to all that God had commanded him, so he did. ... Then Noah built an altar to the Lord, and took of every clean animal and of every clean bird and offered burnt offerings on the altar.*
>
> – Genesis 6:14, 22; 8:20

Hands can make music ... can make food ... can make tools ... can make art ... or build a house ... or erect an altar.

One of my most cherished memories of our recent visit to Barbados was at the world-famous *"Earthworks"* pottery shop, watching the hands of the master craftsman, David Speicer, take a slab of rather ugly clay, spin it quickly on the potter's wheel, and creatively raise up a beautiful pot or vase, or a piece of artwork that is as functional as it is lovely. Later, the shapely but drab pot is painted with a muddy gray color, which – when subjected to the heat of the furnace – blooms into radiant Caribbean blues and greens. (Lots of spiritual parallels there, don't you agree?)

3. Hands make war or peace:

> *And when Abram heard that his relative had been taken captive, he led out his trained men, born in his house, three hundred and eighteen, and went in pursuit as far as Dan. He divided his forces against them by night, he and his servants, and defeated them; and pursued them as far as Hobah, which is north of Damascus. And he brought back all the goods, and also brought back his relative Lot with his possessions, and also the women and the people.*
>
> – Genesis 14:14-16

Unfortunately, Humanity has a long history of aggression, of making treaties of peace but quickly breaking them ... including God's own people, who entered into covenant with Jehovah (see Genesis 14:1-4, 7, 9-10), and then turned their backs on Him again and again. But look at the word-picture: a man coming into agreement with another will look face-to-face and eye-to-eye with the other, extending his bare, open hand to seal the bargain. Yet the man breaking an agreement will look away, turn his back, withdraw his hand, or even take up a weapon. In our relationship with God, do we see ourselves as "givers" and "agree-ers" or as "takers" and "breakers"?

4. Hands can comfort and nourish:

Now the Lord appeared to Abraham by the Oaks of Mamre, while he was sitting in the tent door in the heat of the day. And when he lifted up his eyes and looked, behold, three men were standing opposite him; and when he saw them, he ran from the tent door to meet them, and bowed himself to the earth, and said, "My lord, if now I have found favor in your sight, please do not pass your servant by. Please let a little water be brought and wash your feet, and rest yourselves under the tree; and I will bring a piece of bread, that you may refresh yourselves; after that you may go on, since you have visited your servant."

And they said, "So do, as you have said."

So Abraham hurried into the tent of Sarah, and said, "Quickly, prepare three measures of fine flour, knead it, and make bread cakes." Abraham also ran to the herd, and took a tender and choice calf, and give it to the servant, and he hurried to prepare it.

And he took curds and milk and the calf which he had prepared, and placed it before them; and he was standing by them under the tree as they ate.

– Genesis 18:1-8

My paternal grandmother, Lily Jane Rogers Gossett, made the best fruit and cream pies! As a little girl, I watched with fascination as she quickly mixed flour, water, shortening, and salt to form a dough, then deftly kneaded it, rolled it out, filled the pie pan ... and created the lightest, tenderest pies filled with delectable fruits and flavors.

(Sadly, somehow that knack never made it to her granddaughter; although I've watched and studied and **tried,** my pie crusts have always been tough and disappointing. I've tried many different recipes and had other "experts" instruct me, but I've never matched the excellence of my grandmother's pies ... and I should just stick to baking cookies and cakes.)

5. Hands can protect and rescue:

Now the two angels came to Sodom in the evening as Lot was sitting in the gate of Sodom. When Lot saw them, he rose to meet them and bowed down with his face to the ground. And he said, "Now behold, my lords, please turn aside into your servant's house, and spend the night, and wash your feet; then you may rise early and go on your way." ... Yet he urged them strongly, so they turned aside to him and entered his house; and he prepared a feast for them, and baked unleavened bread, and they ate.

Before they lay down, the men of the city, the men of Sodom, surrounded the house, both young and old, all the people from every quarter; and they called to Lot and said to him, "Where are the men who came to you tonight? Bring them out to us that we may have relations with them."

But Lot went out to them at the doorway, and shut the door behind him, and said, "Please, my brothers, do not act wickedly." ...

But they said, "Stand aside." ... So they pressed hard against Lot and came near to break the door. But the angels reached out their hands and brought Lot into the house with them, and shut the door. ...

Then the angels said to Lot, "Whom else have you here? A son-in-law, and your sons, and your daughters, and whomever you have in the city, bring them out of this place; for we are about to destroy this place, because their outcry has become so great before the Lord that the Lord has sent us to destroy it."

And Lot went out and spoke to his sons-in-law, who were to marry his daughters, and said, "Up, get out of this place, for the Lord will destroy the city." But he appeared to his sons-in-law to be jesting. ...

So the angels seized Lot's hand and the hand of his wife and the hands of his two daughters, for the compassion of the Lord was upon him; and they brought him out, and put him outside the city.

– Genesis 19:1-2, 3-7, 9-10, 12-14, 16; paraphrased

I watched an old World War Two movie recently, where the badly out-numbered squadron of allied forces was preparing for a last ditch effort to overcome the enemy. For several minutes of film-time, with a catchy tune of crisp military music playing, the cameras showed soldiers preparing their weapons: clipping in magazines of bullets, snapping together rifles and guns, securing grenades and small bombs to their uniform pockets and belt-loops, clapping on helmets and goggles and binoculars, clattering and zipping and tying and slapping with purpose and authority – and probably with a lot of extra sound effect enhancement. It built confidence that the good guys were going to save the day.

Next I switched to an international cable news channel showing a real-life scene of Emergency Medical Technicians helping people climb out of a recent disastrous bus-wreck. Desperate, blood-smeared hands reached up out of the tangled vehicle to be caught up by the hands of strong, ready, trained workers, who were trying to save lives. There are different ways to protect and rescue, using hands.

6. Hands offer sacrifices to God:

> *Then they came to the places of which God had told him, and Abraham built the altar there, and arranged the wood, and bound his son Isaac and laid him on the altar on top of the wood. And Abraham stretched out his hand and took the knife to slay his son.*
>
> *But the angel of the Lord called to him from Heaven and said, "Abraham! Abraham!"*
>
> *And he said, "Here I am."*
>
> *And he said, "Do not stretch out your hand against the lad, and do nothing to him; for now I know that you fear God, since you have not withheld your son, your only son, from Me."*
>
> *Then Abraham raised his eyes and looked, and behold, behind him a ram caught in the thicket by his horns; and Abraham went and took the ram, and offered him up for a burnt offering in the place of his son.*
>
> – Genesis 22:9-13

Here is where our responsibility as *"kings and priests"* (see Revelation 1:6) is revealed. Giving our seed in sacrifice has always been God's place of salvation. Note that Abraham *"raised his eyes and looked"* (verse 13) to discover the "answer" which God had **already** placed within his reach.

Keep reading through the Book of Genesis, and you'll discover many ways your hands can be used by God.

***

HANDS
By Don Gossett

Ordinary hands, empowered by God, can be used to minister healing, deliverance and impart rich blessing. Lifted toward God in prayer and praise, hands become immensely important to Him.

I conducted a gigantic open-air crusade at Georgetown, St. Vincent, in the West Indies. Night after night, I stood on the platform and saw hundreds of black hands raised in surrender to the Lordship of Jesus. One night, a police officer was sent from the government to interrogate the crusade chairman, Pastor Charles, about the ministry. In particular, he was concerned about the "peculiar manner" of worship by the lifting up of hands in praise, prayer and worship to God.

"Sir," I addressed the officer, "we know that Jesus said, *'God is a Spirit, and they that worship Him must worship Him in spirit and truth'* [John 4:24; KJV]. The burning of candles, erecting statues, crawling on hands and knees have been Man's way of approaching God. But the Bible commands, *'Enter into His gates with thanksgiving and into His courts with praise'* [Psalm 100:4; KJV). This is why we worship by lifting up of hands, because God's Word is truth and His truth dictates the lifting up of our hands."

I shared with the officer that uplifted hands are the universal sign of surrender, as well as the indication of victory acknowledged. "Where did all this begin?" he asked.

"As far as I know," I replied, "it began with a man named Moses, in the Bible. God commanded Moses to lift up his hands while Israel was in a particular battle. As long as Moses' hands were lifted to God, Israel prevailed. When his hands dropped to his sides, the battle was reversed against Israel. Two men, Aaron and Hur, then supported Moses' hands in the air, and Israel won the battle." When I showed the policeman the Biblical reference in Exodus 17:11-13, he was satisfied and departed.

Throughout the Bible there is a chain of Scripture indicating the significance of men's hands in praise and worship. *"I will therefore that men pray everywhere, lifting up holy hands, without wrath and doubting"* (1 Timothy 2:8; KJV). Many Christians ignore this New Testament command, but it is a joy to obey and lift up holy hands to God.

"Because Thy loving kindness is better than life, my lips shall praise Thee; thus will I bless Thee while I live; I will lift up my hands in Thy Name" (Psalm 63:3-4; KJV). Praising the Lord *"with all that is within us"* includes the beautiful lifting up of our hands to the Lord!

"Lift up your hands in the sanctuary and bless the Lord" (Psalm 134:2; KJV). In public worship, all things are to be *"done decently and in order"* (1 Corinthians 14:40; KVJ). But there is absolutely nothing indecent or disorderly about praising Jesus *"in the sanctuary"* by lifting up of hands!

"Let my prayer be set forth before Thee as incense; and the lifting up of my hands as the evening sacrifice" (Psalm 141:2; KJV). The Jews were required to offer a

lamb as *"the evening sacrifice."* Now, through Jesus, we are to offer the sacrifice of praise unto God continually (see Hebrews 13:15). The "lifting up of our hands" in praise is as the evening sacrifice, so pleasing to the Lord.

"I stretch forth my hands unto Thee; and my soul thirsteth after Thee, as a thirsty land" (Psalm 143:6; KJV). The stretching forth of our hands is an indication of our wholehearted and earnest thirsting after the Lord Jesus Christ, Who alone can satisfy our thirst.

Worshipping Hands

While serving as a coordinator at the first *World Conference on the Holy Spirit* in Jerusalem, one of my delights was to stand in the balcony and watch the thousands of delegates from all over the world worship the Lord in the universal expression of the lifting up of hands. The uttering of words of praise were in many languages, but the expression of lifting up of hands in thirst after the Living God was the same!

The lifting up of holy hands in prayer, praise and worship is God's desire of us. Man's order in religion is often ceremonial, ritualistic, after Man's tradition. But God's order in worship is life, liberty and the lovely lifting up of our hands.

Not only are our hands to be used in worship and praise, but in ministry as well. With my family of seven, for many years we have joined hands while saying Grace around the dinner table. This beautiful function of touching each other draws us closer together. If you seek to employ touch as much as possible, you will discover it's a vital communicator of Jesus' love. There is something "magic" about being in touch.

Christians ought to embrace each other often. We take our example in the importance of touch from our Master. He used this method to communicate love to little children: *"And He laid His hands on them"* (Matthew 19:15; KJV). And we read in Mark 10:16: *"And He took them up in His arms, put His hands on them, and blessed them."* The tender touching of children is very important and so Scriptural!

Your hands can be vitally used in God's service. Hebrews 6:2 speaks of the *"doctrine of laying on of hands."* Hands can be laid upon the sick, and Jesus promised them complete recovery. Hands can be laid on for the baptism in the Holy Spirit, the impartation of spiritual gifts and various Word-oriented ministries.

The laying on of hands is more than a ritual – it imparts a divine charge of the Spirit of God. *"And the Lord said unto Moses, 'Take thee Joshua, the son of Nun, a man in whom is the Spirit, and lay thine hands upon him; and set him before Eleazar the priest, and before all the congregation; and give him a charge in their sight'"* (Numbers 27:18-19; KJV). *"And Joshua, the son of Nun, was full of the Spirit of wisdom; for Moses had laid his hands upon him"* (Deuteronomy 34:9; KJV).

Confirmation

The laying on of hands is the Bible method of impartation of gifts and the confirmation of callings. *"And they chose Stephen, a man full of faith and of the Holy Ghost, and Philip ... whom they set before the Apostles; and when they had prayed, they laid their hands on them"* (Acts 6:5-6; KJV).

The love of Jesus in your hands is also the means of ministering healing to the sick. Of Christ it is written, *"He laid His hands on every one of them, and healed them"* (Luke 4:40; KJV). *"And He laid His hands on her; and immediately she was made straight and glorified God"* (Luke 13:13; KJV). The Apostle Paul also ministered in this fashion: *"Paul laid his hands on him, and healed him"* (Acts 28:8; KJV).

If we have compassion for the sick and suffering, what shall we do? Jesus knew the love in our hearts would compel us also to desire to minister healing to the sick in His Name, so He promised us: *"These signs shall follow them that believe ... they shall lay hands on the sick, and they shall recover"* (Mark 16:17-18; KJV). With love in your hands, seek to lay those hands upon the sick as often as possible.

The Holy Ghost baptism is also ministered by the laying on of hands: *"Then laid they their hands on them, and they received the Holy Ghost. And when Simon saw that through the laying on of the apostles' hands the Holy Ghost was given, he offered them money"* (Acts 8:17-18; KJV). Indeed, there is great value in the laying on of anointed hands; but it's the gift of God, not to be purchased with money. *"And when Paul laid his hands upon them, the Holy Ghost came upon them, and they spake with tongues and prophesied"* (Acts 19:6; KJV). Your hands of love become channels of power through which you can minister the baptism in the Holy Spirit.

Also, you can lay on your hands in sending forth Gospel workers to witness: *"When they had fasted and prayed, and laid their hands on them, they sent them away"* (Acts 13:3; KJV). Before you lay on hands there is a specific Spirit-led directive and caution which we're commanded to observe: *"Lay hands suddenly on no man"* (1 Timothy 5:22; KJV).

You are a Jesus person. Christ lives in your heart and His love is shed abroad within you by the Holy Spirit. Now your identification as a Jesus person is His supreme love that flows within your spirit. You love with His love.

Your hands may be ordinary, perhaps have warts and calluses on them. But those ordinary hands empowered by God can be used to minister healing, deliverance and impart rich blessing. Lifted toward the Lord in prayer and praise, your hands become immensely important to God.

Thank God for your hands! Let them be instruments of good and blessing.

Chapter 9: **FEAR NOT!**

When I was born with severe respiratory problems (as I've previously related in this book), God healed me. I was too young to remember anything about that time, except for still having faint scars on my feet. But when I was 5 or 6 years old, I contracted tuberculosis. My maternal grandfather, William Canada Shackelford, is the one who noted I was pale, skinny, coughed a lot – and since several members of his own family had died of tuberculosis during the Depression years, he recognized my symptoms.

I did not know there was something wrong with me. I knew I coughed too much, and always seemed to have a runny nose. I often had trouble breathing. I was restricted from playing outdoors with my brothers and sisters, but since I was an avid reader, I actually enjoyed the long hours of sitting in the window seat, watching my siblings at boisterous play outdoors while I was quietly indoors devouring book after book.

Perhaps I contracted the illness from a classmate, because in the late 1950s public school system in Tulsa, Oklahoma, we were participating in a "great experiment" of desegregation: we were in classes with dozens of little black children straight from the ghetto. My best friend in First Grade at the school in Tulsa was a little black boy named Patrick – who always seemed to have a runny nose and a higher-than-normal temperature.

I remember often putting my head under my pillow at night, so my constant coughing would not wake up the rest of the family. But then I usually got in trouble the next morning because my coughing also provoked the spitting up of blood ... and blood on my pillow and sheets was not a sign easily hidden.

Following Grandpa Shackelford's recommendation, my parents took me to our family doctor, Dr. Rex Graham, who examined me. He put this odd-looking "patch" on my skin in the middle of my back – some kind of diagnostic test to determine the presence of tubercular germs. As a young child, I wasn't told all the reasons why, but I took very seriously the restriction: "Don't get the patch wet!" I sense there was something pivotal about this "patch" on my back, and how, as a young child, I would be required to use my own faith to receive divine healing.

Well do I remember the day when God healed me of tuberculosis. It happened while walking through the city streets of downtown Chicago, Illinois, where my father was conducting meetings at the church of Pastor A.T. Smith. Our families were spending a day in a park, the children playing, the adults talking. But then Dad and Pastor Smith stopped right in the middle of the sidewalk, called me over to them, and began to pray for my healing. Pastor Smith put his huge black hands on my skinny little shoulders, and I staggered, struggling to remain upright under the weight and force of his hands, and the shaking and swaying he did as he prayed. But I also remember that I chose to actively join my faith to the adults – mainly because I wanted to go play in the park with all the other kids!

Then God healed me. I just knew it. I sensed His strength pouring through the hands of Pastor Smith and my father. If strength can be gentle and thorough simultaneously, that would be how I described that healing power. I just knew God had healed me of tuberculosis.

When we returned to Oklahoma after the Chicago meetings were over (about two weeks later), Mother and I went to see Dr. Graham. He pulled the "patch" off my back. He left the examining room for awhile, consulting with other medical personnel, then he returned. "Now, Jeanne," he sternly asked, "did you get this patch wet?"

I solemnly and truthfully replied, "No, sir." I genuinely meant it.

Then he thoughtfully explained to my mother, "These test results are contradictory. It appears that an indicator of the presence of tubercular germs **started** to appear, then faded away. It's hard to say ..." The disease had been in me ... then it went away. No treatment, no medicine, no human "intervention" – just God's healing power through the big black hands of Pastor A.T. Smith and the big white hands of my Daddy.

After that time, I gradually stopped having prolonged coughing spells, I never again spit up blood, and I eventually gained weight and strength. So I knew I was healed. A few months later, when my family immigrated to Canada, I was subjected to a battery of physical tests, including testing for tuberculosis – and I was passed with a "clean bill of health."

Side-Effect

There was, however, a side-effect to this experience with tuberculosis which I, as a little girl, "acquired": I became very afraid of suffocating and/or drowning. Perhaps there were subconscious reasons – such as remembering the experience of sputum and blood filling my lungs – but I was hesitant to learn to swim because I carried within my heart a certain fear of water and a definite fear of drowning.

There was a more than physical victory achieved on the day I learned to hold my breath and swim underwater – there was something being conquered in my spirit too. This fear of drowning continued to plague me throughout my life, often manifesting itself in my dreams (which have always been extraordinarily vivid).

In my early teens, one summer a family with teenaged children came to visit our family. We all went swimming at the local public pool. Their son – whom I'll call "Wally" – thought he was being quite funny when he playfully held my head under the water. I don't know where the pool's life-guard on duty was, but we are all playing down at the deep end of the pool when I became the target of Wally's joking around.

I was just holding onto the edge of the pool, minding my own business, when, for no apparent reason, Wally pushed me away from the edge and into the deep water. Holding my hair in his fist, he kept my head under water for several seconds. In my surprise, I hadn't taken a very deep breath prior to submerging, so when he finally let me up, I was gasping and very frightened. This was no joke to me – but Wally did it again, and again, and again! Although much smaller and lighter than he, I clawed at his hands and fought with all my strength from under the water. Perhaps it was the blaze of anger in my eyes that Wally saw on that last time, but quite abruptly he stopped tormenting me and let me go.

I was furious! At first I just fought to regain my breath, but when I pilled myself out of the water, I launched into a tearful, angry verbal attack on Wally that probably blistered his skin as well as his ears. Fueled by the fear of drowning now turned into anger, I screamed and sputtered my outrage at this boy. I didn't care about making a spectacle of myself at the public pool – he had nearly killed me, and I wasn't going to slink away in shame. I yelled at this kid, probably embarrassing him more than anything else. For some reason, Wally's family never, ever came to visit with us again.

To this day, although I've had a couple bouts with pneumonia, I am a decent swimmer. I can enjoy a day on a boat at a lake, and have taken sea-cruises in *very* deep water ... but I continue to identify with that old familiar spirit of the fear of drowning. The fear of drowning occasionally slips into my dreams, creating another opportunity for spiritual warfare, even at night. I recognize that suffocatingly evil spirit – and I rise up against it with righteous anger!

God Will Make A Way

(Verse 1) *God will make a way where there seems to be no way*
He works in ways we cannot see, He will make a way for me
He will be my Guide, hold me closely to His side
With love and strength for each new day
He will make a way, He will make a way

(Bridge) *By the roadway in the wilderness, He'll lead me*
And rivers in the desert will I see
Heaven and Earth may fade
But God's Word will still remain
He will do something new today

(Verse 2) *God will make a way where there seems to be no way*
He cares for you, He understands, He holds you in His loving hands

He will do for you what no other friend can do
With love and strength for each new day
He will make a way, He will make a way.

– Don Moen and Jeanne Halsey

One Last Story

With my lengthy preface to my younger years, I want to relate one last Bible story that deals with allowing God to show you the way out of a frightening, potentially deadly situation. Perhaps it comes as no surprise that one of my all-time favorite Bible stories has to do with the fear of drowning. That event of not-coming-up-for-air, that experience of being overwhelmed and lost in darkness, that sensation of perishing without hope – that *"hanging onto the end of my rope"* which so plagues people today – that is a fear with which I am all-too familiar.

Like other new Testament accounts, this same story appears in three of the four Gospels: Matthew, Mark and Luke.

Matthew's Eye-Witness Account

And when He got into the boat, His disciples followed Him. And behold, there arose a great storm in the sea, so that the boat was covered with the waves; but He Himself was asleep.

And they came to Him, and awoke Him, saying, "Save us, Lord; we are perishing!"

And He said to them, "Why are you timid, you men of little faith?" Then He arose, and rebuked the winds and the sea; and it became perfectly calm.

And the men marveled, saying, "What kind of Man is this, that even the winds and the sea obey Him?"

– Matthew 8:23-27

Mark's Version

And on that day, when evening had come, He said to them, "Let us go over to the other side." And leaving the multitude, they took Him along with them, just as He was, in the boat; and other boats were with Him.

And there arose a fierce gale of wind, and the waves were breaking over the boat so much that the boat was already filling up. And He Himself was in the stern, asleep on a cushion; and they awoke Him and said to Him, "Teacher, do You not care that we are perishing?"

And being aroused, He rebuked the wind and said to the sea, "Hush, be still." And the wind died down and it became perfectly calm. And He said to them, "Why are you so timid? How it is that you have no faith?"

And they became very much afraid and said to one another, "Who then is this, that even the wind and the sea obey Him?"

– Mark 4:35-41

The Way Luke Heard It

Now it came about on one of those days, that He and His disciples got into a boat, and He said to them, "Let us go over to the other side of the lake." And they launched out. But as they were sailing along, He fell asleep; and a fierce gale of wind descended upon the lake, and they began to be swamped and to be in danger.

And they came to Him and woke Him up, saying, "Master, Master, we are perishing!"

And being aroused, He rebuked the wind and the surging waves, and they stopped, and it became calm. And He said to them, "Where is your faith?"

And they were fearful and amazed, saying to one another, "Who then is this, that He commands even the winds and the water, and they obey Him?"

– Luke 8:22-25

Those Silly Disciples!

What variety within a single story! I see a lot of finger-pointing going on here. The disciples felt like it was Jesus' fault that they were afraid. Although they knew He had been pouring out Himself in ministry to the crowds of people and could really use the rest (He must have been utterly exhausted to sleep through such a storm), yet their fear of drowning overwhelmed them and they importuned the Master. Notice they were not motivated by faith in His ability to save them – they were motivated by fear.

The Storm

I've thought a lot about the nature of that storm. Do you think it was one of Satan's devices to try to destroy the Son of God? He **had** tried several times before (see King Herod's nefarious activities in Matthew 2, for example). Or do you think the storm was yet another opportunity for Jesus to glorify God (see Jesus' response to the death of Lazarus in John 11)? Do you think the physical elements – the wind, the water, even the buoyancy of the fishing boat –

were **not** subject to Him?

Who then is this, that He commands even the winds and the water, and they obey Him?

– Luke 8:25

In the beginning was the Word, and the Word was with God, and the Word was God. He was in the beginning with God. All things came into being by Him, and apart from Him nothing came into being that has come into being.

– John 1:1-3

I think the storm would have become worse if the disciples had not finally wakened the Master – but I don't think they would have drowned:

"Look! I see four men loosed and walking about in the midst of the fire without harm, and the appearance of the Fourth is like the Son of God!"

– Daniel 3:25

Jesus spoke to them, saying, "Take courage, it is I; do not be afraid."

And Peter answered Him and said, "Lord, if it is You, command me to come to You on the water."

And He said, "Come!" And Peter got out of the boat, and walked on the water and came to Jesus.

– Matthew 15:26-29

Do you think Jesus was really asleep? Actually, I do think He was sleeping because, in this case, Jesus was operating entirely as "God in flesh" ... and I know that intensive ministry can really take it out of you.

And a woman who had a hemorrhage for twelve years, and could not be healed by anyone, came up behind Him, and touched the fringe of His cloak; and immediately her hemorrhage stopped. And Jesus said, "Who is the one who touched Me?"

And while they were all denying it, Peter said, "Master, the multitudes are crowding and pressing upon You."

But Jesus said, "Someone did touch Me, for I was aware that power had gone out of Me."

And when the woman saw that she had not escaped notice, she came trembling and fell down before Him, and declared in the presence of all the people the reason why she had touched Him, and how she had been immediately healed. And He said to her, "Daughter, your faith has made you well; go in peace."

– Luke 8:43-48

Jesus was exhausted ... but He was not deaf.

It's Your Choice

Jesus not only rebuked the storm, He rebuked the disciples for lacking faith. His ability to control the elements – after all, He was the Son of the Creator of the entire Universe! – was not the pivot of this story. If I were God (aren't you glad I am not), I don't think I would have given Mankind the "right to choose." Man's free-will ability to choose his own path, his own response, his own destiny, is one of the most sensitive and troublesome aspects of our relationship with God. But He clearly gives us free-will:

See, I am setting before you today a blessing and a

curse; the blessing, if you listen to the commandments of the Lord your God, which I am commanding you today; and the curse, if you do not listen to the commandments of the Lord your God, but turn aside from the way which I am commanding you today; by following your gods which you have not known.

– Deuteronomy 11:26-28

Now, therefore, fear the Lord and serve Him in sincerity and truth; and put away the gods which your fathers served beyond the river and in Egypt, and serve the Lord. And if it is disagreeable in your sight to serve the Lord, choose for yourselves today whom you will serve. ... But as for me and my house, we will serve the Lord.

– Joshua 24:14-15

So many people think they are totally in control of their lives. But the Bible clearly teaches that either we follow God ... or we follow that other guy. Contemporary folk-rock singer Bob Dylan put it this way: *"You got to serve somebody."* Whether you choose to believe and obey God and His commands in the Word, or choose to "believe not" – which is making a default decision to believe the lies that Satan has been feeding you – ultimately you **do** have to serve somebody! This, too, is falling into one of the traps for our faith which the enemy sets.

God in His wisdom gave Man the right to choose ... and because we're so susceptible to the tricks of the enemy, Man generally make the wrong choices. In this case, the disciples chose selfishness, fear and panic. **However,** because of God's great love for we His children, we have this enduring promise:

I will never leave you, nor will I forsake you.

– Hebrews 13:5; also see Joshua 1:5

This is the pivot of the story! Jesus was in the boat with the disciples ... and the Spirit of Jesus Christ is with you at every moment of your life. When you think you're about to drown, call on Him! When you think the end of that hopeless rope is about to unravel and you will fall into the eternal abyss, call on Him! Come to Him with faith and confidence, not doubt and unbelief. Come to Him with expectancy and appreciation, not hopelessness and despair. God honors faith – and He rebukes timidity!

Look to Him as your total Source, and He in turn will show you what you already have in your hands that will expedite your salvation. Then exercise the practice of giving thanks **before** the provision is manifested.

Summary

What's that you have in your hands? As stated, I believe our loving Heavenly Father, Who has a purpose for each of our lives, will not fail to equip us with all the tools, all the gifts, all the opportunities needed to fulfill our destinies. I pray you will earnestly seek the Lord, will find Him in *"the burning bush,"* will listen to His commands, believing for the *"impossible"* (as did the Widow of Zarephath), and see your seed of "potential abundance" turn into overflowing blessings – even twelve baskets full! – and bring great glory to God as you go about *"walking and leaping and praising God."* Amen!

BIBLIOGRAPHY

The New American Standard Bible, Open Bible Expanded Edition; published 1985 by Thomas Nelson, Inc., La Habra, California, U.S.A.; all rights reserved; used by acknowledgement and consent.

The Argument by Jeanne M. Halsey; previously published in *"Three Strikes";* copyright 1996 by ReJoyce Books, a division of Triumph Communications, Inc.; Blaine, Washington, U.S.A.; all rights reserved; used by permission.

"We Seek Your Face" by The Parachute Band; copyright 1998 by Parachute Band Music; all rights reserved; used by acknowledgement and consent.

"That's When" by Helena Barrington; copyright 1993 by Integrity's Praise! Music/BMI); all rights reserved; used by acknowledgement and consent.

"Michelangelo – Complete Edition" by Ludwig Goldscheider; copyright 1975 by Phaidon Press Ltd.; London, England; distributed by Praeger Publishers, Inc.; New York, New York, U.S.A.; all rights reserved; used by acknowledgement and consent.

"Through Heaven's Eyes" from *"The Prince of Egypt";* lyrics by Stephen Schwarz, music by Hans Zimmer; copyright 1998 by Dreamworks Pictures LLC.; all rights reserved; used with acknowledgement and consent.

"Hindered Prayers" by Jeanne M. Halsey; previously published in *"The Ministry of Drama";* copyright 1997 by ReJoyce Books, a division of Triumph Communications,

Inc.; Blaine, Washington, U.S.A.; all rights reserved; used by permission.

"An Illustration of Faith" by Jeanne M. Halsey; previously published in *"Three Strikes"*; copyright 1996 by ReJoyce Books, a division of Triumph Communications, Inc.; Blaine, Washington, U.S.A.; all rights reserved; used by permission.

"I Expect a Miracle"; chorus by Judy A. Gossett; copyright 1994 by Songs of Triumph/ASCAP; verses by Jeanne M. Halsey; copyright 1999 by Songs of Triumph/ASCAP; all rights reserved; used by permission.

"I Want to Drink From the River of Life"; chorus by Ty Tyler; copyright information unknown; verses by Jeanne M. Halsey; copyright 1999 by Songs of Triumph/ASCAP; all rights reserved; used by permission.

"He Knows My Name" by Tommy Walker; copyright information unknown.

"Hands" by Don E. Gossett; copyright 1972 by Bold Bible Missions; Blaine, Washington, U.S.A.; all rights reserved; used by acknowledgement and consent.

"God Will Make a Way"; verse 1 and bridge by Don Moen; copyright information unknown; verse 2 by Jeanne M. Halsey; copyright 1994 by Songs of Triumph/ASCAP; all rights reserved; used by permission.

About the Author

Jeanne Halsey is a daughter, sister, wife, mother, grandmother ... and a talented writer. Third of five children born to international missionary-evangelist Dr. Don E. Gossett and the late Joyce Gossett, Jeanne naturally inherited her father's "gift of writing" (he has published over 120 books, including the best-selling *"What You Say Is What You Get"* and the ever-popular *"My Never Again List"*). Jeanne was born in Oklahoma ... immigrated to Canada at age 7 ... was educated in British Columbia (Douglas Junior College, the University of British Columbia) ... and has lived in Oklahoma, Oregon, British Columbia, Washington state, Texas, and Colorado. She has traveled internationally extensively.

Formerly Managing Editor of two internationally-distributed monthly Christian magazines, Jeanne is now a free-lance writer. She has ghosted and published books for several renowned Christian ministries and contemporary personalities: for her father; Robert Tilton; Marilyn Hickey; Danny Ost; Cliff Self; Benny Hinn; Sarah Bowling; Reinhard Bonnke; Paul Overstreet; U. Gary Charlwood; and many others. She has written for Christian and secular trade magazines, and published several Sports articles about NBA superstar Luke Ridnour for *Sports Spectrum* Magazine. She also publishes an Internet newsletter, *e-Jeanne,* and regularly teaches Creative Writing classes.

Jeanne lives in Blaine, Washington, with her husband (since 1974) Kenneth Halsey, a Vice President of Franchise Sales for the *Realogy Corporation;* and their "empty-nest home" includes an American Cocker Spaniel, Maraschino Valentino Royale. Their beautiful daughter Jennifer is married to Patrick Freeman; they have two children, Kristian and Ava; thankfully, they live very nearby in Blaine. Their talented son Alexander is married to Cherry Ruth; they have three children, Jude, Aja and Hayley; they live halfway around the world, stationed in India as missionaries with *Youth With A Mission.*

Jeanne is an outspoken activist for Christian causes, and has run for public office (she lost). She is formerly a member of the Board of Directors for the *Whatcom County Pregnancy Clinic.* Jeanne and Kenneth are active members of *North County Christ the King Community Church* in Lynden, Washington.

Other Titles by Jeanne Halsey

Non-Fiction

Behold the Lamb (an Easter Bible Study)
Three Strikes (Dealing with Apathy, Ingratitude and Unbelief)
As For Me and My House ("How to Win Your Loves Ones to Christ") – with Don Gossett
Born to Conquer – with Don Gossett
The Ministry of Angels and Believers – with Don Gossett
Protect Yourself In a World of Danger – with Don Gossett
Great Transactions of the Power of God – for William Canada Shackelford
Don's Daily Devotions – with Don Gossett
If Nobody Reaches, Nobody Gets Touched – with Don Gossett
The School of Praise – with Don Gossett
The School of Blessing – with Don Gossett
Win the Lost At Any Cost (The Danny Ost Story) – with Harald Bredesen
How to Receive and Keep Your Healing – for Robert Tilton
Break the Generation Curse – for Marilyn Hickey
Through the Bible – for Marilyn Hickey
The Church Alive in Shanghai – for Paul Crawford and Bishop Aloysius in Lu-xian
Courage: How to Make Things Happen – for Cliff Self
Take Your Freedom – for Rita Lecours
My Knight In Shining Armor – with Linda Knight
The End Times Are Over, So What Are You Afraid Of? – with Mike Dearinger and Bob Seymour
What's That You Have In Your Hands?
Solutions – for Sarah Bowling
Fearless On the Edge – for Sarah Bowling
Mark My Word (year-long devotional) – for Reinhard

Bonnke
Forever and Ever, Amen – for Paul Overstreet
The Legacy of Writing (classroom curriculum)
It's Not About Me – for Bryan Duncan
Naked With God
Follow the Yellow Brick Road (workbook) – with Reba Rambo-McGuire and Judy A. Gossett
Training the Human Spirit – for Roge Abergel
In Him – for Randy Gilbert
Unlimited Potential in Christ – for Kim O. Ryan
Falling Out of the Tower
Exit the Dragon: Fierce Faith Meets Modern Medicine
The Mark of the King – with Garth McFadden

Fiction

And God Created Theatre
The Ministry of Drama
The Eye-For-An-Eye-Witness News Skits
Messiah! Bright Morning Star (stage play) – with Reba Rambo-McGuire and Dony McGuire, William & Gloria Gaither, and Judy A. Gossett
Bittersweet (a novel of King David and his first wife Michel)
That Which I Ought to Do (a novel of Paul the Apostle)
Anna the Donkey (a children's Christmas story)
Ya-Ya (a novel)
A Christmas Fantasy for Jude and Ava
An Easter Fantasy for Jude, Ava and Aja
Another Fantasy for Jude, Ava, Aja and Hayley
The Blue Vial (a children's science-fiction trilogy)